DESIGN FOR LEAN SIX SIGMA

DESIGN FOR LEAN SIX SIGMA

A Holistic Approach to Design and Innovation

By

Rajesh Jugulum
Philip Samuel

John Wiley & Sons, Inc.

This book is printed on acid-free paper. ♾

Published by John Wiley & Sons, Inc., Hoboken, New Jersey
Published simultaneously in Canada

For general information about our other products and services, please contact our Customer Care Department within the United States at (800) 762-2974, outside the United States at (317) 572-3993 or fax (317) 572-4002.

Wiley also publishes its books in a variety of electronic formats. Some content that appears in print may not be available in electronic books. For more information about Wiley products, visit our web site at www.wiley.com.

Library of Congress Cataloging-in-Publication Data:

Jugulum, Rajesh.
Design for lean six sigma / Rajesh Jugulum, Philip Samuel.
p. cm.
Includes index.
ISBN 978-0-470-00751-8 (cloth)
1. Six sigma (Quality control standard) 2. Quality control. 3. Production control. I. Samuel, Philip. II. Title.
TS156.J84 2008
658.4′013 – dc22

2007039334

Printed in the United States of America

10 9 8 7 6 5 4 3 2 1

To the memory of my father, J. Gopala Char, who taught me how to be brave and courageous, and to my mother, J. Sarala Bai, and Rekha, Aaroh, and Anush
(Rajesh Jugulum)

Contents

Preface

Every corporation on our planet is on a quest to outperform its rivals in two key business areas – managing its current business for maximizing profit and finding the future of its business for maximizing growth. The objective of the first task is to focus on activities and processes that enable flawless delivery of the promises to customers. The goal of the second task is to identify newer and better promises that will delight customers.

During the latest recessionary years in the United States and around the world, organizations have focused all their efforts on improving processes and gaining cost productivity – in many cases using the Lean Six Sigma approach or strategic redeployment or sourcing. This has led to a phenomenon where, for many companies, the coffers are overflowing, and for the first time in recent years, a near-term retreat from an age of expanding possibilities seems imminent.

Fear not, one might say, growth processes are your friends. Unfortunately, the picture is pretty grim there, too. Although growth is usually one of the top agenda for almost every corporation, most fail to sustain growth on the long term. Several researchers have shown that only about 10 to 15 percent of the companies have managed to outperform their rivals and stay on a growth curve for more than a few years in order to deliver above-average returns to their shareholders.

Usually, there are two paths to deliver growth – mergers and acquisitions (M&A) and organic growth. Both approaches have pretty dismal track records. Researchers from McKinsey, A. T. Kearney, PricewaterhouseCoopers, Boston Consulting Group and *BusinessWeek* have shown that about 60 to 80 percent of the M&A deals worth at least $500 million destroyed shareholder wealth within 18 months of closing. On the organic front, about 70 to 80 percent of the products that established companies put into the market fail.

Historically, growth processes have largely been believed to be at the mercy of random processes. This has generated a sense of risk and unpredictability for the managers responsible for delivering results in a predictable fashion to shareholders. We believe that investing in growth opportunities does not have to be such a gamble. It is through the understanding and leveraging of the key variables responsible for innovation and growth that we can establish a predictable, scalable, and repeatable process. This is the focus of this book.

The book outlines a holistic Design for Lean Six Sigma (DFLSS) approach that businesses can adopt to energize their innovation process, perfect their product introduction process, and improve their new product/service development process. It also addresses the end-to-end DFLSS deployment process with a road map that promises a significantly improved and highly effective methodology for revenue productivity. Besides the road map, the book presents discussions on topics such as capturing voice of the customer (VOC), driving growth for innovation, design for robustness, axiomatic design, and theory of inventive problem solving (TRIZ), as they are considered very important to DFLSS. The book also presents a novel approach for system testing that can be used in validation phase of the DFLSS approach. Several other tools and techniques that can be used in conjunction with DFLSS road map are provided in the book. Much literature already exists concerning these tools and techniques.

In Six Sigma deployment, one of the most challenging aspects is to develop metrics to measure success of the system (it could be process, product, or service related). Quite often, we will come across situations where we have to make decisions based on more than one variable or characteristic of the system. These systems are called *multivariate systems*. Examples of multivariate systems include medical diagnosis systems, manufacturing inspection system, face or pattern recognitions systems, and fire alarm sensor systems. In the last chapter of this book we describe Mahalanobis–Taguchi Strategy (MTS) and its applicability to develop multivariate measurement system. The number of case applications of this method is increasing significantly.

Lean Six Sigma has been successfully applied in a variety of industries and functions to solve business problems and improve performance of products, processes, and services. Designing a product, process, or service to provide the intended function at the lowest cost with Six Sigma quality level (3.4 defects per million opportunities) of value for customers is *Design for Lean Six Sigma* (DFLSS). This book provides a structured, systematic, and disciplined methodology to execute the design process without reducing the importance of the designer's intuition and design experience. This, in turn, strikes a balance between rigor and creativity, resulting in optimal design cycle times.

This book will be helpful for various types of audience:

- Top executives
- DFLSS/ Lean Six Sigma leaders
- Master Black Belts
- Lean masters
- Lean experts
- Black Belts

- Statisticians
- Healthcare experts
- Software designers and testers
- Pattern recognizers
- Banking executives
- Financial market consultants
- Academic community, as DFLSS can be considered as industrial and system engineering activity

The book will help them to accomplish the following:

- Understand DFLSS methodologies, along with various tools, concepts, and principles that are embedded in the DFLSS roadmap
- Successfully deploy DFLSS methodology in a systematic way
- Reduce incidences of fire fighting and time of development
- Bring innovative products in the market
- Maximize return on investment by effectively integrating processes, knowledge, and people
- Become self-sufficient in learning and deployment of DFSS method

There are several books on the subject of Design for Six Sigma (DFSS). We consider this book unique because of structured road map and some of the newer tools and techniques that can be used to successfully deploy DFLSS. If the contents of this book change the mindset of executives, engineers, managers, and other DFLSS/Lean Six Sigma leaders to think differently and apply DFLSS differently, it will have served a fruitful purpose.

Rajesh Jugulum
Philip Samuel

Acknowledgments

Writing the book *Design for Lean Six Sigma* has been challenging and enjoyable. The successful completion of this book would not have been possible without help from many outstanding and talented people.

First, we would like to thank Dr. Genichi Taguchi for his tireless efforts to disseminate robust engineering knowledge by way of teaching, consulting, writing the articles and the books. We are also grateful to him for allowing us to use some of his examples. We believe that robust engineering is essential in DFLSS methodology.

Our special thanks are due to Dr. Nam P Suh, president of KAIST and professor at MIT, for developing axiomatic design theory that benefits the industry and DFSS practitioners in a significant way. We are also thankful to him for permitting us to use some case studies from his books on Axiomatic Design and Theory of Complexity.

Thanks are also due to Dr. Daniel Frey, professor at MIT, for his support and involvement in *robustness through inventions* work that can be considered as part of extended TRIZ methodology.

We are also thankful to Dr. Il Yong Kim of Queen's University, Canada, and Dr. Akira Tezuka of National Institute of Advanced Industrial Science and Technology, Japan, for allowing us to publish the work on micropump robust design.

We also wish to thank Mr. R. C. Sarangi and Professor A. K. Choudhury for their help in PCB drilled-hole quality-improvement study. We would like to thank Dr. Leslie Monplaisir and Mr. Mahfoozulhaq Mian for their involvement in the TRIZ case example presented in the book.

We are grateful to Breakthrough Management Group (BMG) for allowing us to draw from its intellectual property and from overall client experiences.

We are very grateful to John Wiley for giving us an opportunity to publish this book. We are particularly thankful to our editor, Mr. Bob Argentieri, for his continued cooperation and support from the beginning. He has been quite patient and flexible with us in accommodating our requests. We would also like to thank Mr. Daniel Magers, the editorial assistant, and Ms. Nancy Cintron, production editor for their continued cooperation and support in this effort.

Finally, we would like to thank our families, especially Lisa Samuel and Rekha Jugulum, for their understanding and support throughout this effort.

Chapter 1

Introduction

This introductory chapter defines the goals of this book with a structured approach to successfully deploy Design for Lean Six Sigma (DFLSS) in any organization. The chapter introduces DFLSS road map that will form a basis for the entire book. The chapter also provides summaries of various chapters in the book and their relation to with each other. It also provides a discussion on the differences between this book and other DFSS books in the market.

1.1 THE GOAL

The goal of this book is to outline a holistic DFLSS approach that businesses can adopt to energize their innovation process, perfect their product introduction process, and improve their new product/service development process. This book will also address the end-to-end DFLSS deployment process with a road map that promises a significantly improved and highly effective methodology compared to other problem solving and design approaches.

Six Sigma has been successfully applied in a variety of industries and functions to solve business problems and improve performance of products, processes, and services. Designing a product, process, or service to provide the intended function at the lowest cost with Six Sigma quality level (3.4 defects per million opportunities) is referred to as Design for Six Sigma (DFSS). This book focuses on the application of the Design for Lean Six Sigma (DFLSS) methodology to achieve desired intent. The approach we provide is a structured, systematic, and disciplined methodology to execute the design process without reducing the importance of designer's intuition and design experience. This, in turn, strikes a balance between rigor and creativity, resulting in optimal design cycle times. The proposed strategy will ensure the following:

- Market driven designs, preventing overdesigns
- Fast, reliable, and more predictable development times
- Focused innovation and inventiveness for growth
- Measurable design activity
- Traceable design logic
- Quick and effective design upgrades
- Robust and reliable designs
- Minimized complexity
- Flexible and modular design
- Designs or redesigns at a six-sigma level

1.2 DESIGN FOR SIX SIGMA – STATE OF THE ART

There are several DFSS books in the market. We have conducted a thorough review of existing books. These books provide a basic overview of DFSS, and some present a collection of tools that will aid DFSS process with very good descriptions and examples. What is missing in these books – and what this book aims to provide – is a structured road map with descriptions of various concepts, tools, and techniques in both engineering and service context. *Design for Lean Six Sigma* (DFLSS) provides illustrations and case studies with real-life examples so that the reader can easily understand and utilize the concepts. Since we are combining DFSS method with lean principles, our methodology referred to as Design for Lean Six Sigma (DFLSS).

We consider that this book stands on its own on the following points:

- This book integrates various concepts, principles, and tools in a unique way for the successful deployment of DFLSS.
- This book offers DFLSS methodology with examples from both service and manufacturing applications.
- This book helps practitioners understand the DFLSS road map in systematic way with emphasis on practical examples and case studies.

It is to be noted that although techniques like quality function deployment (QFD), failure mode and effect analyses (FMEA) and pugh concept selection are very important in DFLSS process, we are not providing descriptions on these topics since there is extensive literature available on these topics.

1.3 APPROACH

The overall approach to meeting the goal of this book is developed based on the road map to deploy DFLSS in all type of industries across the globe. In our road map, DFLSS processes are synchronized and are aimed at helping the organizations design processes in a systematic and meaningful way. This totally modular DFLSS approach allows a flexible methodology that is adaptable to every existing or new design process, independent of the model chosen or used in the organizations.

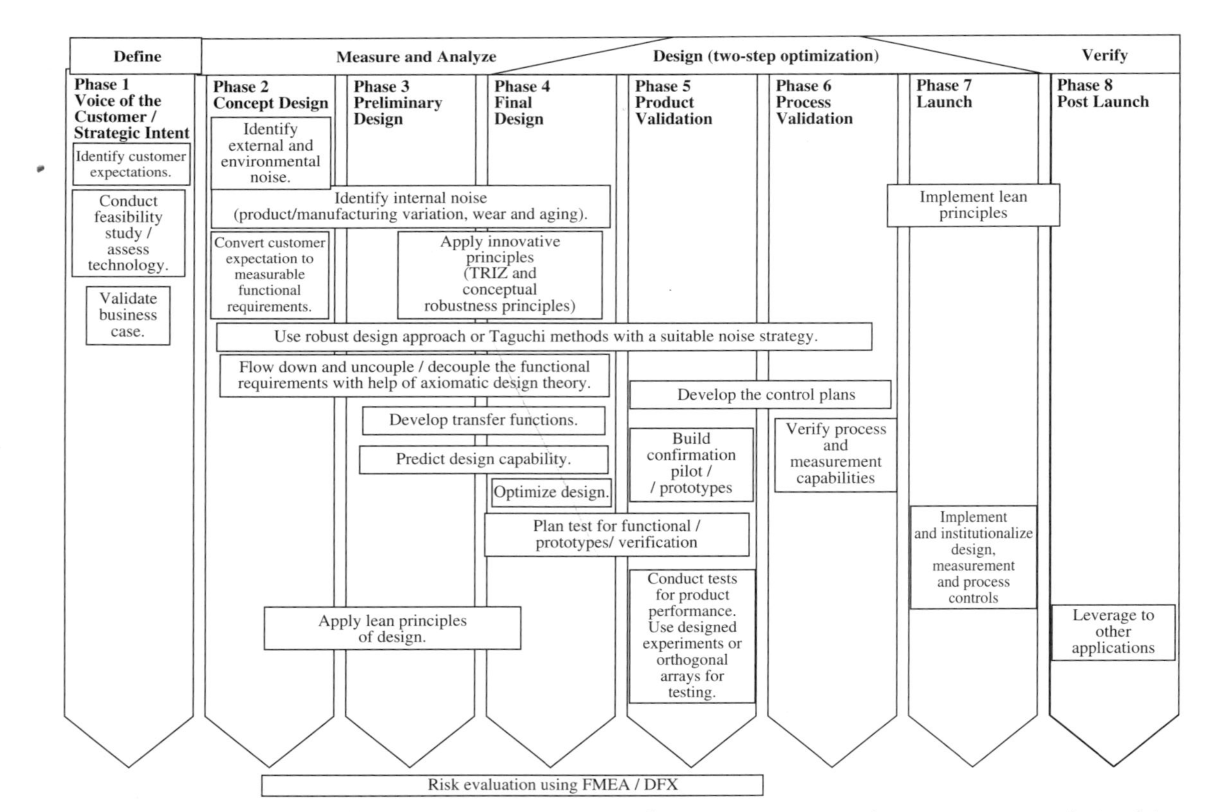

Figure 1.1 DFLSS Roadmap using DMADV (Define, Measure, Analyze, Design, and Verify) Approach.

Figure 1.1 gives various phases of the DFLSS methodology. Various engineering, managerial and statistical concepts, tools, and philosophies are integrated and used in a systematic fashion to achieve Six Sigma quality levels. This DFLSS road map is built in accordance with DMADV (define, measure, analyze, design, and verify) methodology. It is an eight-phase approach that covers all the requirements of DMADV and is aligned to the following main steps:

1. **Define** – Identify customer needs and strategic intent.
2. **Measure and analyze** – Deliver the detailed design by evaluating various design alternatives.
3. **Design** – Design from a productivity (business requirements) and quality point of view (customer requirements), and realize it.
4. **Verify** – Pilot the design, update as needed and prepare to launch the new design.

Phase 1: Customer expectations – In this phase, the customer expectating are identified. After this step, a feasibility study needs to be conducted, and the business case is validated. This corresponds to the *define* step in the DMADV approach.

Phase 2: Concept design – Customer expectations are converted to actionable and measurable functional requirements. Functional requirements are also referred to as critical quality characteristic (CTQs). In this phase, the expectations are flowed to lower levels to understand the design requirements better and come up with good concepts. Application of robust design approach through identification internal (inner) and external (outer) noise will also takes place in this step.

Phase 3: Preliminary design – A detailed design is identified by evaluating various design alternatives. In this phase also, flow down approach and robust design strategies will be used. Pugh concept selection process may be used to select the best alternative against required criteria such as cost or cycle time.

Phases 2 and 3 correspond to the *measure* and *analyze* steps of the DMADV approach.

Phase 4: Final design – The design from a productivity (business requirements) and quality point of view (customer requirements) is developed. Development of transfer function, Taguchi methods of robust design, and two-step optimization are very important in this phase for the purpose of optimization. Readers can refer to Chapter 10 for Taguchi methods and two-step optimization. The capability of the design is also predicted in this phase.

Phase 5: Product validation – Final design is tested against predicted performance and capability. A pilot design is created and confirmation run is conducted to validate the performance of the design. Once this is done, a control plan will be put in place to sustain the gains.

Phase 6: Process validation – The process that was used to build the product is measured and verified, and an appropriate control plan is selected for process variables.

Phase 7: Product launch – The final design is brought to actual practice and results of earlier phases are implemented. The design, measurement controls, and process controls are institutionalized.

Phases 5 to 7 correspond to the *design* step of the DMADV approach.

Phase 8: Postlaunch – The results of this product design are verified and if possible leveraged with other applications. This step corresponds to the *verify* step of the DMADV approach.

1.4 GUIDE TO THIS BOOK

The following brief descriptions of the various chapters will help readers quickly browse through the contents of this book.

Chapter 1 is an introductory chapter that provides goal of this book, an approach for DFLSS deployment with the help of a road map and other distinctive features of this book.

Chapter 2 highlights the role of innovation for growth and survival in the market place. The focus of creating the future is to drive growth and achieve long-term objectives. It also describes the role of management of such organization for successfully integrating and separating cultures, processes, systems, and structures that operate at opposing levels.

Chapter 3 provides a process for systematic innovation by focusing on the evolution aspect of identifying and creating newer and better promises to the customers through innovations resulting in new and better-performing products.

Chapter 4 gives a brief introduction to Lean Six Sigma methodology, including *Define, Measure, Analyze, Improve, and Control* (DMAIC) phases, the concept of variation, and the concept of lean designs. The chapter also briefly highlights the importance of various tools, techniques, philosophies, and concepts of Six Sigma and principles of elimination of waste.

Chapter 5 explores how we might create a system across the organization to deploy the principles of Design for Lean Six Sigma (DFLSS). It also talks about creating an infrastructure and establish a governance structure that promotes the deployment objective if the principles of DFLSS were to become pervasive and the processes of DFLSS were to be followed rigorously.

Chapter 6 describes how we capture *voice of the customer* (VOC) by understanding customer needs and mapping them into functional domain, and then create products or service in such way that these needs are met flawlessly, thus providing value for the customers. This chapter is very important to DFLSS methodology.

Chapter 7 is dedicated for *axiomatic design*. This theory helps us map VOC into the functional domain; map functional domain to physical domain; and map physical domain to process domain to enable us to follow a process to flawlessly design a product or service to satisfy customer needs. The chapter describes two axioms related to this subject: independence axiom and information axiom with examples.

Chapter 8 is dedicated to implementing lean design strategies and related approaches that are required for DFLSS. These strategies aim to maximize value and minimize waste. The chapter also gives brief introduction to 3P process, as popularized by Toyota.

Chapter 9 provides a discussion on the theory of inventive problem solving (TRIZ) methodology, with a systematic road map and examples. A distinctive feature of this chapter is that we have included an add-on section to TRIZ in the form of *robustness through inventions*. This section gives about nineteen principles that would be helpful to create robust concepts. These principles are classified by using the elements of parameter diagram, or p-diagram.

Chapter 10 gives an overview for design for robustness-based Taguchi approach for robust engineering. Robustness concepts are very important for DFLSS, and strategies for countering noise effects can help increase market share of a company in this competitive world. This chapter describes different aspects of quality, strategies for reducing effects of noise factors, role of signal-to-noise ratios, and importance of simulations in the design for robustness. Real-world case studies are used as examples to illustrate concept of robustness.

Chapter 11 provides a new method for testing a system (product or service) by using designed experiments or orthogonal arrays. It helps in testing the product under various customer usage conditions and studying two-factor combination effects, in addition to main factor effects. The methodology is explained with the help of case studies and is useful in later stages of DFLSS methodology.

Chapter 12 discusses the development of multivariate measurement system using the *Mahalanobis–Taguchi Strategy* (MTS). MTS applies in situations where we have to make decisions based on more than one variable or characteristic of the system. These systems are known as *multidimensional systems*. This method will be helpful in creating a measurement system in multidimensional cases and thus is in line with Six Sigma–based thinking, where we always talk about metrics and challenges associated with measuring success of the system with a higher degree of confidence.

Chapter 2

Driving Growth through Innovation

To succeed in the marketplace, every company must master two sets of activities – manage the present and create the future. The purpose of managing the present is to focus on the short-term objectives and generate profits. The focus of creating the future is to drive growth and achieve long-term objectives. In order to successfully navigate the bridge between short- and long-term objectives, companies must engage in a third set of activities known as *selectively abandoning the past*. However, there are only very few organizations who are adept at all of these activities over a period of time.

2.1 DELIVERING ON THE PROMISE

Every company makes a promise to its customers in order to fulfill certain customer needs. In response to the promise, customers generate certain expectations regarding the promise. Often, the experience customers have with the company, from the first contact to the delivery and support of the product or service, is strikingly different from the original promise. Managing the present and improving the performance of current business is all about flawlessly delivering on the promise made to the customers. Performance improvement approaches such as Lean Six Sigma provide a framework for successfully identifying and closing the gaps between expected performance and actual delivery.

Unfortunately, a large number of companies proficient at delivering on the promise are deficient at pioneering breakthrough innovations. Many of these companies use Lean Six Sigma or other management systems to manage, measure, and improve their existing systems and processes. The cultures of these companies thrive at *doing things better* and perfecting the current paradigm. The climate at these companies is one of stability, risk averseness, and orientation toward details and efficiency.

At one point in history, Kodak mastered the analog photography business and was known for excellence in producing high-quality films and printing papers for photography. However, it stayed too long in the analog photography paradigms and was too slow to adopt digital photography as a new means to satisfying unmet customer needs. Digital photography has already altered the basis for competition within the industry and virtually rendered the analog products obsolete. As a result, Kodak has lost its market leadership within the industry and is left to play catch-up.

2.2 CREATING A BETTER PROMISE

Although delivering flawlessly on the promise made to the customers is important for profit generation and outperforming the rivals in the present state, it is not a guarantee about the future growth and profitability of the company. Customer expectations are constantly changing, while competitors and new entrants are placing new bets and thus creating new promises. Therefore, in order to deliver shareholder value and stay on the growth curve, companies must explore new paradigms, create new promises, and deliver them flawlessly to the customers ahead of the competition.

Identifying opportunities and creating new promises requires a different set of skills and competencies than the ones required for perfecting existing systems and paradigms. This is where innovation capabilities play a key role in accelerating growth and securing the future. Again, we find that many companies who thrive on innovation do not have an environment or culture that enables them to perfect their systems and deliver flawlessly.

These companies are adept at *doing things differently* and exploring for new opportunities and ways to satisfy customer expectations. They are typically known for embracing change, dissatisfaction with status quo, risk taking, and less regard for conventional wisdom or rule. The characteristic and style of people and culture good at innovation are at odds with the ones who like to perfect the existing system and stay with the current paradigm.

Xerox Corporation and its Palo Alto Research Center (PARC) provide an interesting example of a company that was adept at pioneering inventions but failed to capitalize on its innovations. They were well known for many breakthrough ideas such as laser printers, modern PC GUI paradigm, ethernet, and object-oriented programming. Xerox has been often criticized for its inability to commercialize and profitably exploit many of PARC's innovations. The culture and climate at PARC was favorable for exploration and pioneering work. But they were largely unable to perfect their innovations and exploit them by successfully taking it to the marketplace.

2.3 AMBIDEXTROUS ORGANIZATION

Companies that stay on the growth curve for the long term and deliver above-average return to shareholders must learn to be adept at both sets of activities – improving and managing the present as well as continually creating the new future. In other words, these companies are ambidextrous by simultaneously delivering flawlessly on the promises to the customers while discovering and creating new and better promises. The key task of business leadership is to guide the balancing activities that will optimize against satisfying existing customers with current offerings and that will sustain the growth objectives for the long term.

Apple is an example of a company that does well in managing the paradoxes associated with preservation and evolution. While Apple continues to execute flawlessly on the design, manufacture, and marketing of personal computer and associated software and peripherals, it also continuously innovates on the line of entertainment and communications solutions such as portable digital music players, as well as related accessories and services, including online sale of third-party audio and video products. Apple has demonstrated its masterful skill at product and business model innovation through iPod and iTunes and continues to demonstrate mastery at product, store, and experience design. Now it is extending its reach to the living room and cell-phone market. In order to continue the winning streak on growth and deliver the profitability to shareholders, Apple cannot waver on perfecting its existing offerings to its customers while looking for new opportunities to extend its reach.

Delivering on current promises to existing clients in the form of customer solutions provides an entry-level ticket for companies to compete. It's no longer a long-run competitive advantage. Product life cycles are getting shorter, customer expectations are changing, and technology and globalization are rewriting the basis for competition. Therefore, companies have to mine new sources for differentiation and competition.

Several researchers have identified the difficulty associated with staying on growth curves. For example, Foster and Kaplan (2001) described the phenomena of the rise and fall of corporations included in *Forbes* 100 and S&P 500 with their growth battle. Chris Zook and James Allen describes the similar phenomenon from a sample of 1,854 companies (Zook and Allen, 2001). Jim Collins also revealed his conclusions in "Good to Great," in which he described his studies of the growth performance of 1,435 companies for the period of 1965 to 1995.

To celebrate the seventieth anniversary, *Forbes* published its *Forbes* 100 list of largest U.S. companies and then compared it to the original list from 1917. Interestingly, only eighteen of the original companies managed to stay in the "Top 100" list through 1987. Sixty-one of them no longer existed, and twenty-one fell off the "Top 100" list. The overall long-term return to the shareholders from this group of eighteen companies was 20 percent less than that of the overall market. Only two companies, General Electric and Kodak, performed better than the overall market. Since then, Kodak's performance has deteriorated significantly.

Another analysis of the S&P 500 revealed that during 1957 to 1998, only seventy-four of the original five hundred remained on the list. Only twelve of the seventy-four performed better than the S&P 500 index (Foster and Kaplan, 2001).

Jim Collins's study of the performance of 1,435 large corporations during the period of 1965 to 1995 showed that only 9 percent of these companies outperformed for a decade or more. Similarly, Zook and Allen study found that only 10 percent of the 1,854 companies they studied grew consistently over a period of ten years.

These studies and several others attest to the difficulty associated with maintaining a high growth rate over a long period of time. Why is it that most corporations with excellent management processes and controls perform poorly compared to the aggregate capital markets and the indices that represent them? A key part of the answer is in the fact that capital markets are built on the assumption of managing the paradox of ambidextrous. The capital market rewards the players who are successful at generating *discontinuity*, thus breaking away from its peers and the efficient operation that leads to profitability and short-term success. In the early 1900s, the Austrian economist Joseph Schumpeter described this phenomenon and called it the process of *creative destruction*.

In the short term, corporations are rewarded for efficiently and flawlessly delivering on the promises made to their customers. However, in order to stay on the growth curve, they must find new sources of competitive advantage and may have to abandon the very promises that made them successful in the short term. In the modern times, the pace of discontinuity has dramatically increased. This is evident from the turnover rate of corporations from the indices that represent the capital markets. For example, in the early 1990s, the turnover rate of corporations who were members of S&P 90 were less than 2 percent. This rate has begun to reach double-digit percentages in recent times.

Let's take the example of companies that have been very successful in remaining on the high growth curve in more recent times (say, for the last ten years). Companies such as Apple, Google, Microsoft, Toyota, GE, P&G, and 3M come to our attention. On one hand, these companies are well known for their operational efficiency and delivering on their promises to the customers. On the other hand, they have been very successful at creating discontinuities. While Apple is busy maintaining its reputation for producing easy-to-use and elegant desktops and laptops, it is also busy finding new promises for satisfying jobs to be done, such as iPod, iTunes, and iPhone. In many scenarios, creating discontinuities involves abandoning the company's existing line of business or cannibalizing present products and services it delivers to the customers.

2.4 PLATFORMS FOR GROWTH

There are two primary platforms for growth – the organic approach, and mergers and acquisitions. Neither approach has a very impressive track record. A study published by *Business-Week* in conjunction with Boston Consulting Group examined a thousand deals worth at least $500 million between 1995 and 2001 and concluded that 61 percent of the deals destroyed shareholder wealth in the process. Their data showed that buyers in these deals repeatedly made a list of common mistakes such as overpaying, overestimating synergies, trouble integrating operations of the merged companies, overemphasizing cost cutting, and underemphasizing revenue maintenance and the retention of top salespeople.

Results on the organic front are not much better, either. Approximately 75 percent of the new products introduced by established companies fail (Leonard-Barton, 1995). Another, but related, study indicates that approximately 80 percent of all venture capital investments fail. However, venture capital firms are typically well aware of this difficulty and hence hedge their bets accordingly.

Given the difficulties associated with sustaining growth, most companies use a combination of organic growth and mergers and acquisitions to achieve shareholder expectations. Therefore, growth through organic approach is on the top of the agenda for executives responsible for delivering results. Innovation and design is the key approach through which firms drive organic growth. The objective of innovation is to create new value for customers. The new value for the customer can come in the form of new or enhanced products, services, or business models. Also, new value can be generated for the business and/or customer from behind the scenes processes.

2.5 INNOVATION AND DESIGN

Primarily, there are three types of innovations – product or service innovation, process innovation, and business model innovation. When we create a new product or service for customers so that they can achieve certain outcomes in a superior way through its use, it is called *product* or *service innovation*. For example, the mobile phone was a product innovation when introduced, as customers were now able to communicate remotely from places with no access to land telephone lines. *Process innovation* involves creating new value in many behind-the-scenes process levels. For example, a number of years ago, FedEx created a system that allows customers to track the location of their packages during transit and delivery. Although customers only saw the tracking information on the Internet and the scanning operations performed by the driver, most of the innovation occurred in the enabling business processes. This created value for FedEx as well as the customers. A *business model innovation* involves delivering superior value by changing the way business is done. Wal-Mart is well known for its business model innovation in logistics and supply chain area. Apple's iPod is a product innovation, while iTunes is a business model innovation. The iTunes business model enabled various music providers and users to come together and share music that created better value for the users, providers, and Apple.

The first phase of innovation, regardless of the three types, is the *ideation phase*. The ideation phase is where we generate a new idea for product, process, or business model innovation. The ideation happens when an unmet customer need and a certain capability (or technology) meet. Bringing the idea alive in the form of a product (or service), process, or business model is the function of design.

Therefore, we can say that design is the bridge between the ideation process and fulfillment process. This can be explained with the help of value chain processes depicted in Figure 2.1. This figure shows how a product or service is created and delivered to customers. An idea is first generated against an unmet need in the marketplace. The idea is possible because we have a certain technology or capability that will enable the creation and delivery of the idea into a feasible product. The idea is perceived as an interplay of various dynamics in the marketplace, including competitor offerings, and customer needs as well as technology and other capabilities. The idea is further developed as a design of the product to be offered to the customer. The design is translated into a solution that is ready for customers via the production and supply processes.

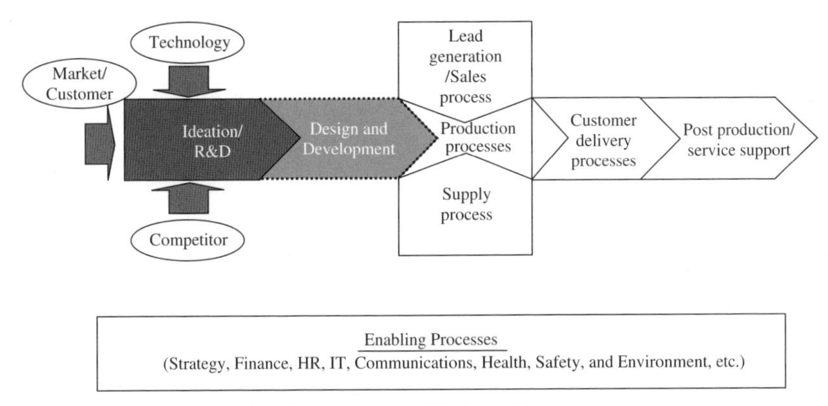

Figure 2.1 Key Value Processes within a Company.

Meanwhile, through sales and marketing processes, we create a promise to the potential customers. This promise is regarding the product or service we will deliver to the customer. Once the customer buys the promise, we now will deliver the product or services through certain delivery and support processes. The value chain demonstrates two major activities. The first activity is regarding the creation of a new and superior promise for the customer. This is achieved through the ideation and design process. The second activity is regarding the fulfillment of the promise through production and delivery processes. The design plays a key role as the bridge between the identification of the new promise and flawless delivery of that promise.

Therefore, if the design is not done correctly, we don't have the right blueprint for the new promise that we have developed. If the design is not right, we may not be able to deliver flawlessly on the promise. We may have to rework or improve the design in order to deliver better on our promise.

2.5.1 Managing the Paradox of Preservation and Evolution

From the previous discussions, it should now be clear that long-term profitability and growth depend on mastering both preservation and evolution capabilities. On one hand, we must be adept at delivering on the promises we made to our customers in order to be successful in the short term and generate profits. This includes excelling and improving at all the processes associated with sales, supply, production, delivery, and support processes. On the other hand, we must be successful at identifying new paradigms and creating newer and better promises. Many times, while the first set of activities are focused at improving an existing promise, the second set of activities are focused on creating a newer promise that would replace the first promise.

In addition, the set of structures, culture, and systems that enable the first set of activities are at odds with the second set of activities. This creates the tension between the two sets of activities.

So the key challenge for companies is to figure out how to create a culture, system, and process to become ambidextrous. In exploring this further, one recognizes that the companies that have been successful in delivering flawlessly on their promises while identifying and creating new and better promises have figured out a way to embrace and thrive in the paradox of structure – it is enabling and limiting at the same time.

Companies who are adept at *doing thing better* or exploitation have taken advantage of the enabling aspects of structure. They have created structures and systems that measure, manage, and improve their existing systems, processes, and paradigms. For example, Lean Six Sigma principles place a heavy emphasis on defining, measuring, and analyzing the existing system in order to make improvements or perfect the system.

Organizations with an innovative climate have recognized the limiting nature of structures and hence embraced a culture and systems that enable change, risk taking, and challenging existing rules or conventional wisdom. A few but growing list of companies such as Apple, 3M, Procter & Gamble, General Electric, and Toyota have found a way to embrace and simultaneously implement and manage this paradox and thrive in it. They have adopted a diverse cognitive climate of style, level, and organizational structure that enables the paradox.

2.5.2 Types of Paradoxes

So what are some of the types of paradoxes that ambidextrous organizations must embrace and incorporate? While there are many, we have identified several below based on our experience.

Culture

In ambidextrous organizations *doing things better* (also called adaptive) and *doing things different* (also called innovative) cultures coexists. The adaptive side of the organization embraces efficiency. Doing it right the first time is important for this side. Everything is done methodically, and mastering the details is a very important aspect of this culture. On one hand, this approach is very much needed for flawless delivery of the promise. On the other hand, the innovative side of the culture is too eager to change, take risks, and learn fast from failures. This side of the culture has low regard for methodology, structures, and processes. They prefer to find a new system rather than perfect an existing system. These cultures often clash, but have learned to live together, leverage each other's strengths, and compensate for each others' weaknesses. This is, in essence, the management of style diversity.

Portfolio

The organization maintains a balanced portfolio ranging from incremental improvements to breakthrough innovations. The adaptive side of the house wants more projects to perfect existing systems and paradigms. These projects are focused on improving the flawless delivery of the promises made to customers. The innovative side of the house wants to develop or place bets on new and potential promises. The ambidextrous organization ensures that the portfolio is dynamically optimized against current and future customer, market, and organizational demands.

Metrics

Ambidextrous organizations must maintain two broad sets of metrics in order to meet the objectives of preservation and evolution. The first set of metrics is aimed at measuring the degree to which we are delivering flawlessly on the promises. Examples of these metrics include profitability; customer satisfaction measures relating to cost, quality and delivery; process performance metrics relating to quality, cycle time, and process capabilities; system performance levels; and people performance measures.

The second set of metrics relates to growth objectives and hence focuses on revenue growth, new market penetration, and innovation. These objectives are achieved through identifying unmet customer outcomes, or *jobs to be done* in unoccupied market spaces. Examples of these metrics include revenue growth from new markets and new products or services; product and service development cycle time; vitality index; quality and strength of innovation pipeline; and percentage of sales from radical innovations introduced by the firm within the last three years.

Often, these two broad sets of metrics are at odds with each other. The first set of metrics often penalizes not doing things right the first time, taking risks, or deviating from the set path. It promotes short-term success. The second set of metrics encourages risk taking, tolerates high failure rates, and rewards long-term success. Ambidextrous organizations know how to incorporate the appropriate metrics for the relevant people, group, and time. It is a balancing act and involves embracing the diversity of objectives and styles.

Process, Methods, Techniques, and Tools

The adaptive side of the organization thrives on achieving objectives systematically and methodically, following certain processes and structures. These set of structures and processes enable efficiency and limit waste and defects. The basic tenets of Lean Six Sigma approach are founded on this premise.

Innovative organizations see the limiting aspects of rigid structures and processes aimed at achieving perfection. But repeatability, scalability, and systematization across the enterprise suffer when innovation is not guided by structure and processes.

Many organizations have found success in the short term from accidental discoveries or inventions from the ingenuity of a single gifted individual or a small cadre of geniuses. However, this approach is not sustainable in the long term. For example, recently *Wall Street Journal* (2007) reported on the fall of Motorola from too much dependence on one innovation – the Razor mobile phone. Organizations that are successful innovators in the long term utilize flexible structures, processes and systems that enable exploration, risk taking, and quest for new paradigms. Just as Lean Six Sigma techniques and tools enable searching inside a paradigm to achieve perfection, there are innovation techniques and tools that enable exploration outside the paradigm in a systematic way.

Simultaneous adoption of preservation and evolution requires the coexistence of adaptive and innovative structures and processes within the organization. Although the operations group within the company might rightly place emphasis on using Lean Six Sigma and the process management approach to reduce variation and eliminate defects, it must also look for opportunities for process innovation to leapfrog the process capabilities, especially where the process has achieved its entitlement. In a similar way, product and service innovation approach can be made more rigorous by borrowing the discipline from adaptive processes and structures.

People

An organization wanting to be successful in both preservation and evolution dimensions over a long period of time must attract, retain, and manage a work force that is diverse in cognitive style and level. It has been shown that humans are born with a preferred problem-solving style that is stable over a long period of time. This preferred style tends to be more adaptive or more innovative on a bipolar scale. More adaptive people prefer to solve problems within the paradigm in tried and proven ways. They enjoy perfecting the system. More innovative people are eager to see the problem from unexpected and tangential angles. They are ready to question the assumptions and beliefs and have less regard for rule or group conformity.

In simple terms, the more adaptive are suitable for most of the preservation activities, while the more innovative prefer to work with activities linked to evolution. In reality, complex problems that companies undertake are solved widely and well by the collaboration of adaptors and innovators. However, adaptors and innovators have trouble communicating, working together, and trusting each other. Leadership of ambidextrous organizations must figure out how to manage the cognitive diversity of style and level needed for the success of the organization.

An additional challenge lays in the difference in which employees are motivated and rewarded for preservation and evolution. Extrinsic rewards are common ways to reward short-term actions that result in flawless execution of current agenda. These types of motivation do not work well for rewarding long-term actions resulting in creation of new and better promises for the customers.

Organizational Structure and Alignment

Every company must create work groups and organizational structures that align well to its short-term mission and long-term strategies. For an ambidextrous organization, this means the parts of the organization responsible for delivering on the current promises must be lean, efficient, and agile in order to execute flawlessly on the promises to its customers. This part of the organization has well-defined, sustaining activities against its short-term priorities. These are further grouped into planned and incremental activities to achieve the cost, quality, and delivery objectives.

By contrast the parts of the organization responsible for achieving evolutionary objectives must have flexible systems and structures that enable successful exploration, risk taking, and entrepreneurial work. This group works well with fewer hierarchies and loose structures with freedom from rigid constraints. The motivation and rewards systems are aimed at optimizing against growth objectives and creating new market spaces in unoccupied territories.

It is the responsibility of the senior leadership to orchestrate the dynamic separation and integration of the teams responsible for preservation and evolution.

WL Gore, a privately held company, is well known for its growth and innovation. Its culture and organizational structure for cultivating innovation is well worth examining. At Gore, there is hardly any hierarchy and very few ranks or titles. The structure promotes direct, unfiltered communications – anyone can speak to anyone else in the company. To make this approach practical, teams are organized into small groups so that members can get to know each other and know what everyone else is working on. At Gore, everyone is your boss and no one is your boss. It has a culture where employees feel free to pursue their own ideas, communicate with one another, and collaborate on their own free will and motivation.

Customer and Market Orientation

Companies focused on preservation know their customer expectations regarding their products or services. This enables them to deliver flawlessly on their promises. The customer expectations about the products or services fall into two categories – performance expectations and perception expectations. *Performance expectations* are objective, unambiguous, and measurable expectations regarding the products or services. *Perception expectations* are subjective, ambiguous, and difficult to measure. However, both expectations are important to the customer.

It is important to note that these expectations change over time, depending on the market dynamics and competitive pressures. Therefore, in order to deliver flawlessly on promises, every company must continually master the details of performance and perception expectations from customers regarding the products or services (the promise).

Customers have a third kind of expectation called *outcome expectations*. These are the reasons that customers are using the product or service. These expectations are related to *jobs to be done*. Identifying and measuring these jobs to be done in unoccupied market space and aligning or developing new solutions for it is the key to successful innovation. Companies that are adept at innovation are masters of market insight and new paradigm exploration.

Therefore, to be adept at preservation and evolution simultaneously, companies must master the details of all three types of expectations – performance, perception, and outcome expectations. This enables ambidextrous organizations to continually deliver on the promises to ever-changing customer expectations.

Many of the products that we use today did not exist three or five years ago. For example, who would have thought that Procter & Gamble would cannibalize its own well-known utilitarian mop in exchange for the Swiffer mop that utilizes electromagnetism to attract and trap dirt? Cirque du Soleil has taken market share away from traditional circus show providers such as Barnum & Bailey through its ultra-high-tech solutions with no elephants and lion tamers at more than five times the traditional ticket price for similar shows. It has done this by mastering the unmet customer outcomes (jobs to be done), inventing new promises, and delivering flawlessly on the new promises.

Leadership

While certain parts of the organization focus on sustaining activities with short-term objectives, others are busy identifying and exploring a new set of activities for evolutionary objectives. Company leadership has a crucial role to play in managing and responding to the different needs of these organizations. While one set of activities are aimed at perfecting existing systems, processes, products, and services, others may be cannibalizing these products or services in favor of better ones.

Therefore, it is the responsibility of the ambidextrous leadership to keep the integrity of the two sets of activities and organization while leveraging, sharing, and learning from each other. The leadership must empower, motivate, and reward teams involved in activities that require styles that are contrary to theirs (adaption versus innovation). Such leaders who can make difficult and objective decisions while working with team members unlike them are rare but essential for managing the paradox of preservation and evolution.

2.6 CONCLUSIONS

Building an organization that outperforms at flawless delivery of the promises made to its customers while constantly creating new promises and selectively abandoning existing promises in order to grow consistently is not an easy challenge. This type of organization must embrace the paradox of preservation and evolution simultaneously. The management of such an organization involves successfully integrating and separating cultures, processes, systems, and structures that operate at opposing levels.

Chapter 3

Process for Systematic Innovation

In the previous chapter, we discussed the general idea of simultaneously pursuing preservation and evolution – flawlessly delivering on the promises while continually identifying newer and better promises and selectively abandoning the past. In this chapter, we take a closer look at the aspect of identifying and creating newer and better promises.

As discussed in the last chapter, market data reveal that staying on the expected growth curve and returning above-average shareholder returns on the long term have become a very difficult task for corporations. In the early 1990s, the turnover rate of corporations who were members of S&P 90 was less than 2 percent. In recent times, this rate has begun to reach double-digit percentages.

Management of innovation and growth processes has generated a sense of high risk and unpredictability for the managers responsible for delivering financial results to shareholders. The variables impacting innovation and growth processes are only beginning to be understood. Many a times, we hear of innovation as a result of happy accidents. Also, we hear of certain individuals being very gifted in idea generation, invention, or innovation. However, if innovation is to become systematic, scalable, and repeatable across the corporation, then we must understand, manage, and leverage the underlying processes that enable innovation.

Let's examine some of these variables that impact the process of innovation:

- A balanced innovation portfolio
- Collaborative, high-performing innovation teams
- A systematic process for executing innovation projects
- Proven innovation techniques and tools
- A climate that supports innovation and creativity
- A governance system to manage the innovation process and activities

3.1 BALANCED INNOVATION PORTFOLIO

Innovation-elite firms understand that achieving uncommon industry growth rates means going beyond the traditional research and development focus. Companies that seek growth through innovation benefit from developing a balanced, comprehensive portfolio that spans many areas – products and services, processes, strategy, and even the organization's core business model. These companies also vary in the required degree of innovation, from incremental to significant to breakthrough levels. (See Figure 3.1.)

An IBM survey of 765 global CEOs in 2006 revealed that although new products and services remain a priority, companies are placing more emphasis on using innovative business models to differentiate themselves from the competition (IBM Global Business Services, 2006). In addition, more and more companies are reaping the benefits of simultaneously going after product, process, and/or business model innovation. Organizations that execute innovation projects in this way almost always

Degrees of Innovation		Types of Innovation		
		Product	Process	Strategy / Business Model
	Breakthrough			
	Significant			
	Incremental			

Figure 3.1 Portfolio of Product, Process and Business Model Innovation.

generate higher return on investment than companies that limit innovation to new products.

For example, Apple has experienced tremendous success with the iPod, a product innovation. However, the success of the iPod is largely due to the introduction of iTunes, a business model innovation. Through this combination of product and business model innovation, Apple created $70 billion in shareholder value in just three years. It also serves as an example of producing high return on R&D investment, since Apple achieved such great results with investments as little as one-tenth the size that Microsoft spend.

The mix, timing, and quantity of portfolio is a function of many factors directed by business strategy. These factors include revenue growth gap, life cycle of existing products and services, fleeting expectations of customers, moves from competitors and availability of new technology, and other capabilities. For example, a consumer electronics company reviewed its revenue growth projection for the next three years. It soon realized that its existing products and services would be expected to produce only 80 percent of the projected growth in revenue for the next three years. This accounted for a projected shortfall of $500 million in revenue growth in the following year. When the next two years were figured in, the projected shortfall reached close to $3 billion. Therefore, the company realized that it needed a new pipeline that would generate $3 billion over the next three years. When the uncertainty and chances of failures were taken into consideration, the company decided to build a portfolio of product, services, and business model innovation that should account for $6 billion to $9 billion over the next three years.

As indicated from the previous example, the timing of portfolio identification must take into account the product or service development cycle time. Development of many ideas often requires the development of certain technology or capability. This can be achieved by developing the capability internally, or licensing it through alliances or joint ventures. But this takes time and, hence, must be orchestrated with the right timing.

Companies can identify and manage their own balanced innovation project portfolio by using a set of growth and innovation opportunity assessment techniques. In addition to project prioritization and scope, these tools help organizations identify unarticulated, latent, and underserved customer expectations that might indicate an unoccupied market space – and a potential direction for growth.

3.2 EFFECTIVE TEAMS FOR COLLABORATION

The next key to building an innovation factory is to assemble innovation teams that are capable of flawless and speedy execution, and then manage these teams for high performance and collaboration. There are at least three factors that we must take into consideration in this context:

1. Motivation of the team and individuals
2. Cognitive level of the team
3. Cognitive style of the individual

Motivation can be influenced by positive and negative reinforcements, as well as personal values and beliefs. Therefore it is important to install an appropriate reward system for innovation teams which will maximize the team motivation. The results of innovation can often be measured only on a long-term basis, and success of innovation is often tied to risk taking and managing uncertainties. Therefore many companies find that the appropriate reward system for encouraging innovation are either intrinsic in nature or provide long-term incentives.

The *cognitive level of the team* involves both the manifest cognitive level and potential cognitive capacity of the team members. The manifest cognitive level is correlated to the knowledge and skill needed to solve the innovation and design problems at hand. The potential cognitive capacity refers to the intellectual capacity that is genetically inherited. This, for example, is measured using means such as IQ. Fortunately, for solving most problems that corporations face today what is more important is the manifest cognitive level.

In addition, another poorly understood factor impacts the performance of teams. This factor is called *cognitive style of the individual*. Unlike manifest level and motivation, cognitive style is a stable characteristic that does not change with age and is considered to be genetically inherited. Cognitive style refers to the modality or way in which individuals solve problems, make decisions, and are creative. People with different styles interact in predictive ways with others. Knowing this and managing it can enhance team performance.

Cognitive style is arranged across a continuum on a bipolar scale, ranging from highly adaptive to highly innovative. The more adaptive individuals are concerned with resolving residual problems created by the current paradigm. Other things being equal, they tend to produce fewer ideas that are more manageable, relevant, sound, and safe for immediate use. They are seen as sound, conforming, safe, and dependable. They expect high success rate from ideas generated. The more innovative, by contrast, search for problems and alternative solutions cutting across current paradigms. They approach problems from unsuspecting angles in a tangential manner. They produce many ideas, seen as exciting, *blue sky*, or *new dawn*. Many depict them as unsound, impractical, and shocking. They tolerate high failure rates.

In the context of methodology in problem solving, the more adaptive approach problems in a precise, reliable, methodological manner. They are, in general, very thorough and pay lots of attention to detail. They welcome change as an improver. They seek solutions to problems in tried and understood ways, with a maximum of stability and continuity. The more innovative approach problems from unsuspected angles, appear undisciplined, and think tangentially. They welcome change as a mold breaker. They enjoy manipulating the problem, querying its basic assumptions.

When it comes to the management of structure, the more adaptive want to maintain continuity, stability, group cohesion, and prudent with authority. They solve problems by use of rule. They are cautious when challenging rules and only do so when they have strong support. They are an authority within given structures. The more innovative, by contrast, are likely to be a catalyst to settled groups and consensual views, and to be radical. They prefer to alter rules to solve problem. They have no problem in challenging rules, with little concern for past customs. They tend to take control in unstructured situations.

Organizations need diverse cognitive styles to solve large complex problems. However, diversity of style can create tension and cause challenges in communications, trust, decision making, and ability to work together. Effective collaboration is based on the successful management of cognitive diversity. It is important to know that each style has advantages and disadvantages in task resolution. An advantage in one situation is a disadvantage in another. Diversity is an advantage for task resolution, but a problem for team management. Knowing each other's style and respecting the strength of the person who has a different style is key to managing diversity.

Cognitive style that is stable differs from cognitive behavior, which is flexible. I *prefer* to behave in my preferred style. I can and do behave out of my preferred style, and this is called *coping behavior*. But coping requires extra energy, and extensive coping behavior may cause stress. An effective manager is adept at matching the style of team members with the type of problem at hand. A great leader will provide an environment where minimum coping is necessary and will receive greatest amount of coping from associates in times of crisis.

As already discussed, assembling the right team that is most suitable for the innovation problem at hand requires optimizing against motivation, cognitive level, and cognitive style of the associates. Toward this end, companies can utilize a set of assessments, inventories, and management approaches to assemble effective and collaborative teams for specific growth projects.

3.3 PROCESS FOR EXECUTING INNOVATION PROJECTS

The next key to make innovation repeatable, predictable, and scalable across the enterprise is to utilize systematic approaches. This means making it systematic using a consistent process that is applied by all teams (as DMAIC is applied by Six Sigma teams, for example). The process must also be robust enough to accommodate multiple innovation pathways; while some growth projects require thinking outside of the box, others require more structure within existing paradigms.

Many models are suitable for the end-to-end aspects of the innovation and design process. The end-to-end process ranges from identifying certain customer needs to generating ideas to developing and demonstrating the solution. Although most of the process models embody similar concepts, the contributions from Wallas and Guilford are worth mentioning.

Graham Wallas (1926) proposed a four-stage process:

1. *Preparation* – defining the problem and exploring the scope and boundaries of the problem
2. *Incubation* – internalizing the problem into the unconscious mind in preparation for idea generation
3. *Illumination* – generating creative ideas, bringing into the consciousness
4. *Verification* – verifying, elaborating, and applying the idea

J. P. Guilford (1950) suggested the distinction between convergent and divergent production (commonly referred to as convergent and divergent thinking) in generation and refinement of novelty. Divergent thinking involves searching for new paradigms that provide multiple solutions and ideas for a problem. Convergent thinking refers to searching inside the paradigm for refinement and narrowing the solution space. The end-to-end innovation and design process involves multiple steps of convergent and divergent phases. The front end of the process involves identifying opportunity for innovation and idea generation. The back end involves developing the idea and bringing it into reality. Design thinking is the back-end process, which is the focus of major portion of this book.

It is common, especially among engineering, manufacturing and high-technology companies to separate the innovation activities from the design and development activities. For example, an airplane manufacturer might generate many new ideas for aircraft wings, select a few of them, design and test the prototype to demonstrate the concept. Once the concept is sufficiently validated, tested and proven, the company might incorporate the new design of the wing into a new aircraft under development. This approach enables minimizing operational risks.

A convenient and efficient way to execute the innovation process involving idea generation and design is to follow a process called D4 methodology, adapted from Wallace (1926). The phases of D4 processes are define, discover, develop and demonstrate. During define phase, our focus would be to identify the job to be done and associated unmet outcomes. This forms the basis of the scope of the innovation project. The objective of the discover phase is to explore various paradigms that will satisfy the unmet outcome expectations for customer and provider. Many ideas generated during the discover phase are narrowed down further in the develop phase. The focus of develop phase is to design the system based on the idea we selected. The design is converted into a prototype or pilot during the demonstrate phase. Much information is gathered during this phase to further refine the design.

Similar to the innovation process, the design and development process can also be broken down further into many phases. Although many models exist, one that is popular involves define, measure, analyze, design and verify (DMADV) phases. The define phase of the design and development process enables us to define and scope the project with clear objectives. Performance and perception expectations from customers regarding the solution are gathered, budget planned, resources allocated and project milestones defined. The focus of measure phase is collect necessary data needed for design activities. Analysis and design phase involves data analysis, concept refinement and selection as well detailed design. During verify phase, we will demonstrate the success of the design through piloting and prototyping. We will explore this model in greater detail in the forthcoming chapters.

3.4 PROVEN TECHNIQUES AND TOOLS

The D4 innovation methodology provides a consistent approach to innovation. D4 practitioners must also understand how to apply a variety of tools and techniques that enable success in each project phase. For example, the main objective of the define phase is to identify unmet customer expectations. Techniques such as ethnography, archetype research, and heuristic redefinition all help capture the unarticulated needs of customers.

The process features tools designed to generate new innovative ideas you can use to meet the unmet needs of your customers. These tools range from random entry techniques to provocation and movement techniques to technical and physical contradictions.

The most promising ideas generated in the discover phase are further investigated during the develop phase using techniques and tools that enable the analysis of data and the subsequent design process. Techniques such as axiomatic design, function structure, conjoint analysis, design of experiments and lean design enable smooth execution through this phase.

Finally, successful solutions are implemented in the demonstrate phase using techniques and tools such as piloting, rapid prototyping and mistake proofing.

The DMADV process described earlier also has a set of techniques and tools that support the efficient execution of the design project. The commonly used techniques and tools are shown in Figure 5.2.

During define and measure phases, techniques such as ethnography, focus groups, surveys, and interview enable us to gather performance and perception expectations from customer.

Measurement system analysis is used to ensure that the data collected is valid, reliable and repeatable. Other techniques such as net present value (NPV), internal rate of return (IRR), work-break-down structure, Gantt Charts, In-out scope and multi-generational plans allow us to manage the project and minimize risk.

The key activities of analyze phase is to develop functional requirements, generate concepts, resolve design conflicts and assess risk. Axiomatic design, TRIZ, design scorecard, design failure modes and effects analysis, and Pugh selection matrix are examples of technique that enable the activities during analyze phase. System, sub-system and component level design are completed during design phase of the project. In addition to the many tools used during the analyze phase, other tools such as simulation tools, conjoint analysis, reliability testing, mistake proofing, modular designs and design of experiments are used in this phase.

Success is demonstrated during the verify phase with the help of piloting, prototyping and feedback data. Many DMAIC techniques such as process and value stream maps, takt time, control charts, measurement system analysis, process capability and standard work are commonly used during this phase. The feedback data is incorporated to improve the design and launch processes.

3.5 CLIMATE FOR INNOVATION

One way to mitigate the challenges of innovation is by establishing a climate that is best suited for innovation; in other words, an organizational culture that promotes calculated risk taking, collaboration, and trust. Such a climate enables people to learn from their mistakes (instead of being punished for them). It also supports quicker execution of ideas and a more agile organizational structure, all of which minimize exposure from innovation risk.

In contrast, the organizational climate that is needed in manufacturing and service delivery is one of certainty, precision and minimization of risk. This type of a climate is necessary to deliver on the promises made to customer and perfect our offering. The focus of our attention in this case is to do things efficiently. However, there are some opportunities for experimentation and process innovation.

The design and development activities serve as the bridge between innovation and manufacturing or service delivery processes. While a great deal of experimentation and risk taking are necessary during design and development, it is also important to get things done efficiently to manage the project risks and milestones.

3.6 THE GOVERNANCE SYSTEM

The importance of a governance system to manage the ambidextrous organization is discussed in previous chapter. On one hand we need to promote entrepreneurship, risk taking and experimentation that enable the organization to identify new opportunities leading to newer and better promises to the customers. On the other hand we need certainty, efficiency, variation reduction, waste elimination and minimization of risk to flawlessly deliver on our promises. It is in this context that we need systems and structures that enable simultaneous management of *doing things efficiently* and *doing things better*. Organization structure, reward and recognition systems, team collaboration approaches, and metrics are some of the key elements that need to be taken into consideration while designing an appropriate governance system.

Chapter 4

Lean Six Sigma Essentials

Corporations around the world attest to the benefits of implementing Lean Six Sigma strategy as demonstrated through its impact in financial savings and customer satisfaction. Lean and Six Sigma philosophy had separate origins. While Six Sigma was started as an approach to reduce operational variation and defects, lean thinking enabled elimination of waste and reduction of cycle time. Six Sigma and Lean are the twin forces that fuel any organization's drive for operational excellence. As they work hand in hand, an organization enjoys their combined benefit on the top and bottom line.

As an integrated strategy, Lean Six Sigma has now become a metaphor for a business excellence system that enables the breakthrough improvement in every part of the organization through process enablement, cost reduction, and increased profits. What is becoming increasingly evident is that Lean Six Sigma is also a multifaceted business management system for achieving and sustaining innovation and revenue growth.

Although Lean Six Sigma originated as a system for improvement purposes, the fundamental tenets and principles behind it can be applied in a proactive manner. Typical Lean Six Sigma is aimed at reducing variation, defects, and waste, as well as improving process speed for existing processes and systems. These principles can be applied in a proactive manner to prevent defects and waste, while minimizing the impact of variation and enabling process speed. This is done in the context of designing and developing processes, products, and systems. This, in essence, is the thinking behind Design for Lean Six Sigma (DFLSS), which is the focus of this book. In the forthcoming chapters we will explore in greater detail how principles of DFLSS are applied. In this chapter we shall discuss the origins and basic principles of Lean and Six Sigma, as well as the integrated approach to Lean Six Sigma (LSS).

4.1 ORIGINS OF SIX SIGMA

If you walked into a bookstore in the streets of London in the late 1700s you would not be likely to find any textbooks on modern business management. However, if you picked up a copy of *Miscellanea Analytica* (London: 1730) by the French mathematician Abraham De Moivre, you would find the roots of a revolutionary management system based on the theory of probability. Later Carl Frederick Gauss (1777–1855) added to De Moivre's great contribution by developing the normal curve as a way to understand probability.

A couple of centuries later, other scientists and business leaders would build on De Moivre's and Gauss's work to find application in business management and hence reap breakthrough financial benefits. One such contribution came from Walter Shewhart, when he showed that three sigma distance from the mean is the point where processes require correction. Many process capability measurement standards such as Cp, Cpk, and ppm (parts per million) defects were later added. Although other noteworthy contributions came from Deming,

Crosby, and Juran, credit for coining the term *Six Sigma* goes to a Motorola engineer named Bill Smith.

Motorola first employed Six Sigma for quality improvement and to gain competitive advantage for its electronic products. AlliedSignal then focused Six Sigma projects on cost reduction to realize over $2 billion in savings in a four-year period. General Electric's spectacular success with Six Sigma in the 1990s convinced other organizations to embrace Six Sigma methodology. Multiple billions of dollars have been saved through the implementation of Six Sigma projects.

Six Sigma was first developed as a statistically based technique to define, measure, analyze, improve, and control (DMAIC) manufacturing processes. To this end, its ultimate performance target is virtually defect-free processes and products (six sigma being the measure of 3.4 or fewer defects per million). Over a period of time, Six Sigma has evolved to become a vision, philosophy, goal, metric, improvement methodology, management system, and customer-centric strategy.

4.2 SIX SIGMA APPROACH

In the field of statistics, *sigma* (σ) represents the standard deviation (a measure of variation) of a population (lowercase s represents an estimate, based on sample). However, this should not be confused with the notion of *sigma level* or *sigma score* of a process. Simplistically, the terms *six sigma process*, *sigma level*, or *sigma value* of a process refer to the idea that if we have six standard deviations between the mean of a process and the specification limit, we will make virtually no items that exceed the specifications limits (see Figure 4.1).

As already noted, the commonly accepted definition of a Six Sigma process is one that produces 3.4 or fewer defects per million opportunities. Statistically speaking, a normally distributed

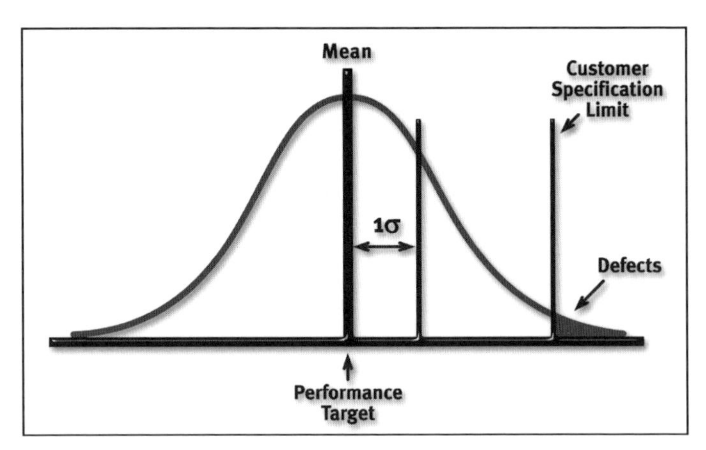

Figure 4.1 Concept of Variation and Sigma Level of a Process.

process will have 3.4 defects per million opportunities beyond 4.5 standard deviations above or below the mean. This would imply that 3.4 parts per million defects correspond to 4.5 sigma and not 6.0 sigma. The 1.5 sigma that is added to 4.5 sigma refers to what is commonly known as *sigma shift*. It is based on the assumption that processes shift and drift over a long period of time. In the absence of specific data for a given process, we assume the shift to be approximately 1.5 sigma level. Therefore, a process capable of performing at 4.5 sigma level on the long term is assumed to be performing at 6 sigma level in the short term.

At the problem-solving level, Six Sigma is a project-driven methodology. Over the course of a typical two-year tenure, trained Six Sigma leaders called *black belts* deploy between eight to twelve high-impact projects that support an organization's overall business objectives. These projects are executed through DMAIC (define, measure, analyze, improve, and control) process. This methodology consists of the following steps:

- **Define** the project with business justification, definition of the defect, problem statement, objective, and project plan.

- **Measure** the current performance of the process collecting the data. Ensure that data are trustworthy using measurement systems analysis.
- **Analyze** the causal relationship between output factor and input factors. Establish the critical factors that have maximum leverage.
- **Improve** the process by optimizing the process using the critical factors.
- **Control** the process and systems to ensure that processes continue to perform at the desired level.

Another important contribution of Six Sigma is reckoning the impact of variation in processes. Variation exists everywhere in nature. No two objects in nature are exactly identical. Therefore, it affects product performance, service quality, and process outputs leading to rework, scrap, and premium freight, all of which can cause customer dissatisfaction. Variation causes uncertainty, risk, and potential defects. There are two types of variation – controlled and uncontrolled. Controlled variation, often referred to as common cause, is a stable or consistent pattern of variation over time (predictable). Uncontrolled variation, referred to as special cause, is a pattern that changes over time (unpredictable). To control and reduce variation we should first understand, quantify, and interpret variation in a data set. The mission of Six Sigma is to identify the areas of variation, isolate root causes, optimize processes and thereby reduce or minimize the impact of variation in our products, processes, and services.

Let us demonstrate the impact of variation with the help of an example. Let us say there are two surgical operating rooms in a hospital performing identical functions. We have decided to study the time it takes to prepare the rooms for surgical operations. Let us say that the average time to prepare the first room is 30 minutes, with high variations resulting in 50 percent

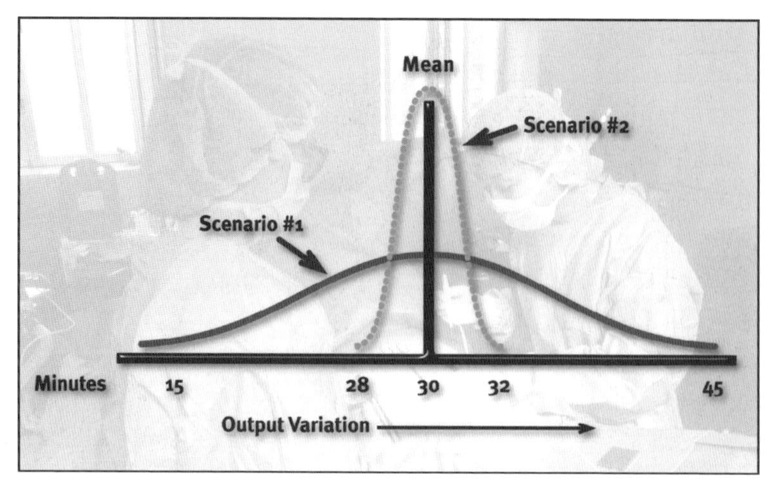

Figure 4.2 Two Processes with Identical Mean but Differing Variations.

of the time preparation procedures taking between 15 and 30 minutes and the other 50 percent preparations taking between 30 and 45 minutes. In the case of second room, average time to prepare is 30 minutes, with little variation resulting in rooms prepped in no more than 32 minutes (see Figure 4.2).

Which scenario is better from certainty, reliability, and risk perspective? Although the average time to prepare is identical, the second room provides consistently reliable support. Uncertainty caused by the first scenario can have a domino effect in downstream operations such as surgery.

Do you remember the last time you waited a long time in a line at the supermarket? Recognize that customers feel the variation and remember it! If it took me 45 minutes one time to receive my service, that is the time I remember, not that it takes 30 minutes on average. It turns out that averages tell us very little about actual customer experience. Customers remember the extremes, not the average. To drive dramatic improvements in performance, the variance in a process must first be minimized.

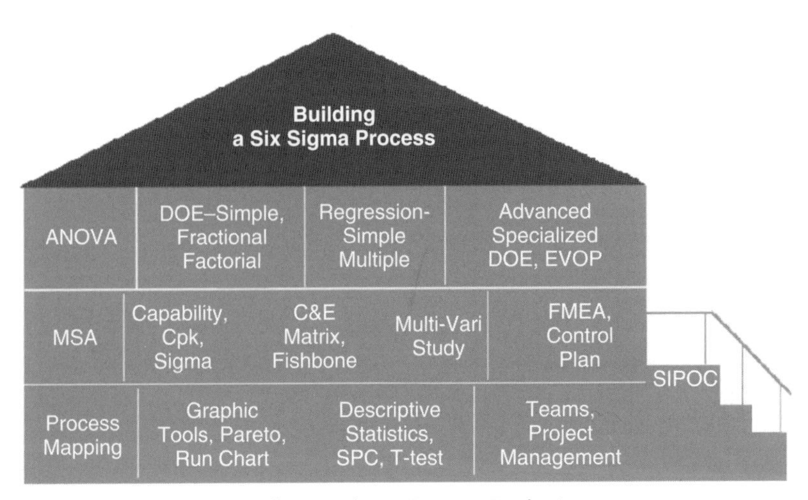

Figure 4.3 Commonly Used Six Sigma Techniques.

The fundamental approach with Six Sigma is embodied in the equation $Y = f(x)$. Y represents the dependent or the output variable we are interested in for improvement. It can be a quality parameter such as reliability, yield, or market share. The x variables are the independent or the input variables that affect the output variable Y. Our objective is to characterize the relationship between Y and x. Once we understand this relationship, we can leverage the key variable to optimize the process and achieve desired level of performance for Y. Six Sigma has a set of techniques that enable us to characterize and optimize the relationship between Y and xs. Many of these techniques are shown in Figure 4.3.

4.3 ORIGINS OF LEAN

Just like Six Sigma, we can trace the roots of Lean back to the 1700s. It all started with Eli Whitney and his inventions of the cotton gin and interchangeable parts for gun. After graduating from Yale in 1792, Mr. Whitney was traveling west of Georgia when he noticed the difficulties growers had with cotton production. Growers had difficulty making money because removing seeds from the fibers required so much time and labor. By 1793, Whitney managed to invent a machine that mechanically separated seeds from fiber. This invention would eventually pave the way to the Industrial Revolution in English cloth manufacturing.

Later on, Whitney would turn his attention to the manufacture of guns. As the need for inexpensive but reliable firearms grew, he saw the potential for mass production using interchangeable parts. He demonstrated this concept in 1798 with the production of ten muskets, which culminated in an order from U.S. military for the manufacture of 10,000 muskets at a price of $13.40 each. Although the production was late due to schedule overrun, he achieved success and sowed the seeds for the industrial revolution led by Henry Ford and others.

The next major round of contributions came from Frederick Taylor (1856 to 1915), Frank Gilbreth (1868 to 1924) and Lillian Gilbreth (1878 to 1972) in laying the foundations for industrial engineering. Taylor pioneered the idea behind time study and standardized work. He called it *scientific management*. Frank implemented motion study and invented process charting, capturing value-added and nonvalue-added elements. Lillian Gilbreth studied psychology and motivations of workers and investigated how attitudes affected the outcome of a process.

Henry Ford (1863 to 1947) is remembered for pioneering mass production, embracing the advances owing to Taylor and others. The hallmark of his contribution points to the River Rouge plant, where he installed a continuously moving assembly line for manufacturing the model T automobile. Ford is considered by many to be the first practitioner of just-in-time and Lean thinking.

After the war, Japanese industrialists studied many U.S. production methods, especially Henry Ford's mass-production system and quality improvement systems advanced by Shewhart, Deming, Juran, and others. At Toyota Motor Company, Taiichi Ohno and Shigeo Shingo would implement many such approaches, including various elements of Ford's mass production system. During a trip to the United States, Ohno was inspired by the Indy 500 racing track, where race cars were getting refueled, tires changed, and serviced in the pit stop in an amazingly short time. Another observation he made was that in U.S. grocery stores, when the inventories were low on the shelf it triggered a replenishment signal to the suppliers. Therefore, material was replenished at the pull of the customers. The true spirit and practice of lean thinking is embodied in the inspirations from River Rouge plant, Indy 500, and U.S. super stores. Taiichi Ohno is regarded as the founder of Toyota production system, the principles of which became later known as lean manufacturing.

James Womack and Daniel Jones (1991) coauthored a book called *The Machine that Changed the World* and coined the term *lean manufacturing*. This was followed up with another book, *Lean Thinking*, (2003). Womack and Jones provided a detailed and straightforward account of the Toyota production system and the associated Lean approach.

Five basic principles of Lean, as explained by Womack and Jones, are as follows:

1. **Value** – Specify value in the eyes of the customer.
2. **Value stream** – Identify all the steps in the value stream and eliminate waste.
3. **Flow** – Allow the value to flow without interruptions.
4. **Pull** – Let the customer pull value from the process.
5. **Continuously improve** in pursuit of perfection.

The main theme behind Lean approach is to improve process speed and reduce cost by eliminating waste. Lean thinking has borrowed heavily from Little's law relating to lead time, work-in-process and average completion rate. It says that when the system is in a steady state,

Lead time = Work-in-process units/Average completion rate

Lead time is the amount of time accumulated between when work entered a process and work leaves the process. *Work-in-process* (or things-in-process) is the quantity of things currently inside the process. *Average completion rate* refers to the number of units processed per unit time. Therefore, the lead time for processing a claim may be calculated by dividing the number of claims-in-process inside the system with average number of claims processed in a given time. In product development, the work-in-process is the number of projects in process. In procurement, the work-in-process is the number of requisitions in process. Lean approach has a set of well-defined tools that can identify opportunities for improving the average completion rate and reducing the work in process.

Lean thinking uses a slightly different approach to problem solving than Six Sigma approach. Progress is made through the execution of events called *Kaizen events*, wherein a small group of employees assemble together to improve certain aspect of the business through a series of quick, focused sessions. This approach reduces the long-cycle-time project mentality and creates a bias toward action. *Kaizen* is a Japanese word for incremental continuous improvement, with everyone working together.

The Kaizen approach for execution is as follows:

1. Map out the current state and create a baseline.
2. Establish a vision for the future state.

3. Identify the gaps and establish opportunities for improvement.
4. Implement changes and remove waste from the system.
5. Evaluate results and institute continuous improvement.

The key thinking behind the Lean approach is to produce what is needed for the customer, when it is needed, with the minimum amount of resources such as materials, equipment, labor, and space. As shown in Figure 4.4, Lean achieves this by attacking eight types of *muda* (Japanese word for "waste"):

1. Waiting
2. Overproduction
3. Rework
4. Motion
5. Processing

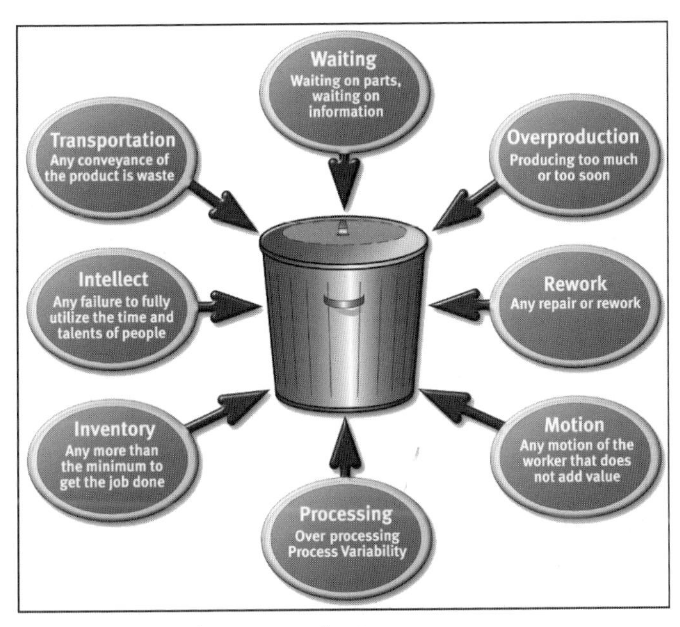

Figure 4.4 Eight Types of Waste.

6. Inventory
7. Intellect
8. Transportation

Waiting

When we stand in line waiting for our turn to receive a service, it is an example of waste. In typical manufacturing processes, more than 90 percent of product's life is spent in waiting to be processed. Much of the time is spent waiting for the next operation. The waste of waiting applies to business processes as well. For example, documents waiting for processing and approval are a non–value-added activity and hence treated as waste.

Overproduction

Producing more than what is necessary is the next type of waste. For example, imagine an automotive supplier produces certain transmission components according to a set schedule. What happens to the products that are produced in excess of customer demand? It takes space, and all those products are tying up cash and holding it idle. In addition, they face the potential of getting damaged or obsolete. In the context of business process, what about a report we have produced that has too much unnecessary information? From a customer's perspective, the customer has to figure out and separate what is useful and what is unnecessary. From the provider's perspective, we have wasted resources and time to prepare and print unnecessary information.

Rework

All the scrap and rework activities are obvious waste. This is straightforward and obvious in the context of manufacturing processes. For example, let us assume that our company manufactures hard disk drives for computers. During the manufacturing process, we may produce many components that might not be fit for assembly into the final product. Alternatively, we might have to rework these components to make them useful for final assembly. These are examples of waste. Rework activities are often called *hidden factory*. Rework is not as obvious in non-manufacturing context. For example, often rework is built into transactional process in the name of *editing*, *approval*, *revision*, and others. Just like in manufacturing processes, our objective should be to get it done right the first time. A flawed strategy or decision can generate much rework for the company.

Motion

Any movement in material that does not change form, function, or fit of the product is a waste from a Lean perspective. Any movement of people or machines that does not contribute to the added value of a product or a service is also a waste. For example, looking for parts, bending/reaching for materials, and searching for tools are examples of wasted motion in the context of manufacturing. Some motion waste can be defined in the context of ergonomically inefficient motion. If an operation creates a repetitive stress injury, this is waste. In transactional processes, often documents are not placed in the most convenient manner or location for processing. This creates unnecessary motion that does not add any value.

Processing

When we process more than necessary, it is considered waste. Engineering change orders typically have many process steps that are not necessary from a customer value-added perspective. It is common for the change orders to go through multiple approvals and sign-offs that are not necessary. In manufacturing, we often perform a certain operation because it has always been done that way.

Inventory

Producing and storing more products, material, parts, or information that is needed to fulfill current customer orders is a waste that falls in this category. In manufacturing context it includes raw materials, work-in-process, and finished goods. Inventory utilizes extra spaces and requires additional handling.

Intellect

This refers to not taking advantage of the thinking power and knowledge base of human resources within the company. Failure to stimulate and capture ideas, not implementing employee suggestions, and poor communications are frequently cited example of waste in this category. Often due to administrative disconnectedness between employees, customer, and suppliers, many opportunities are missed. These, in turn, generate barriers to innovation, efficiency and unnecessary costs.

Transportation

This refers to the unnecessary movement of parts, material or work-in-process from one operation to another. Transportation increases the overall time to process since no value-added activity is carried out during this time. Also, there is a time and resource cost associated with transportation. In addition, damages could occur during transportation. Poorly conceived layout of factory or facility is often the root cause behind this activity. We can eliminate this type of waste by improving the layout, process coordination, housekeeping, and optimization of operations.

Lean approach is to first identify the product family, customer, and the value stream. By mapping out the value stream one can identify non–value-added activities that cause waste. Lean has many techniques that allow us to map out and identify the value stream and the waste. These techniques are used to improve

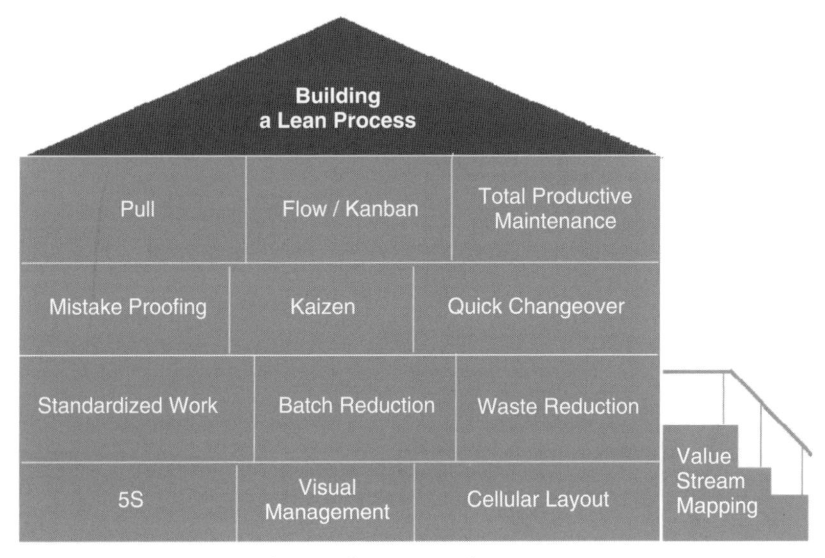

Figure 4.5 Commonly Used Lean Techniques.

the value stream and improve the process speed. Many of the commonly used techniques are shown in Figure 4.5.

4.4 LEAN SIX SIGMA: FASTER, BETTER, AND CHEAPER

For a long time, Lean and Six Sigma approaches were regarded as competing initiatives. Lean advocates noted that Six Sigma does not directly address speed, flow, and waste within processes. Six Sigma supporters pointed out that Lean approach is not capable of solving variation problems or process optimization issues. The logical answer to the dilemma is Lean Six Sigma – the combination of the two world-class approaches to organizational performance. For many corporations, Lean Six Sigma has become an effective operational strategy to be more responsive to changing customer needs, deliver flawlessly on the promises made to the customer, and operate at world-class cost.

On the drive to execute flawlessly on the promises to the customers, corporations constantly strive to accomplish the tasks better, faster, cheaper, safer, and greener. This is true when it comes to operating the processes, developing and producing products, or providing services to customers. The idea behind adoption of the Lean Six Sigma approach is to enable the organization with a common language, framework, methodology, and process to achieve these objectives easily and efficiently.

However, the adoption and blending of the two approaches are not without challenges. On the one hand, when we improve the speed of the process, quality or cost might suffer. On the other hand, when we reduce defects and improve quality, it might increase our costs, reduce the process speed, or degrade the environment. The key to Lean Six Sigma integration is to blend the two methodologies into one approach of getting things done faster, better, cheaper, safer, and greener.

The other challenge of Lean Six Sigma integration stems from the philosophical differences of traditional way of implementing Lean and Six Sigma. For example, Lean is typically implemented through a series of short, focused events called *Kaizen blitz* executed in weeks. Six Sigma is implemented through many projects going though the DMAIC phases that lasts four to eight months. Lean approach looks at end-to-end processes and holds a systems view in order to make improvements. The appropriate analogy is like solving *mile-long-and-inch-deep* problems. Six Sigma projects are typically scoped small, and the approach is efficient for solving complex problems requiring probabilistic and statistical thinking. The analogy here is like solving *inch-long-and-mile-deep* problems.

Therefore, the question is, which problem-solving framework is appropriate – Kaizen or DMAIC? Should the projects be scoped at end-to-end level or scoped narrow? Should we plan on executing projects in weeks or months? Do we need to use all the techniques from Six Sigma and Lean in order to execute a Lean Six Sigma project?

The key to successful integration is to use the right techniques for the given problem and achieve the right performance. Many have found that leveraging the best practices from each approach generates optimum results and performance. For example, combining the rigor of DMAIC problem-solving methodology, quick and focused blitz sessions, and starting with an end-to-end view provides the best of both worlds. Another best practice is to scope the project based on the problem to be solved and use techniques that are appropriate for the problem. A DMAIC framework incorporating both Lean Six Sigma thinking is shown in Figure 4.6.

The integrated approach of Lean Six Sigma optimizes against the value-creation process and maximizes shareholder value by achieving the fastest rate of improvement in cost, quality, process speed, safety, customer satisfaction, invested capital, and environment. We do so by understanding the linkages between financial, customer, process, people, and technology.

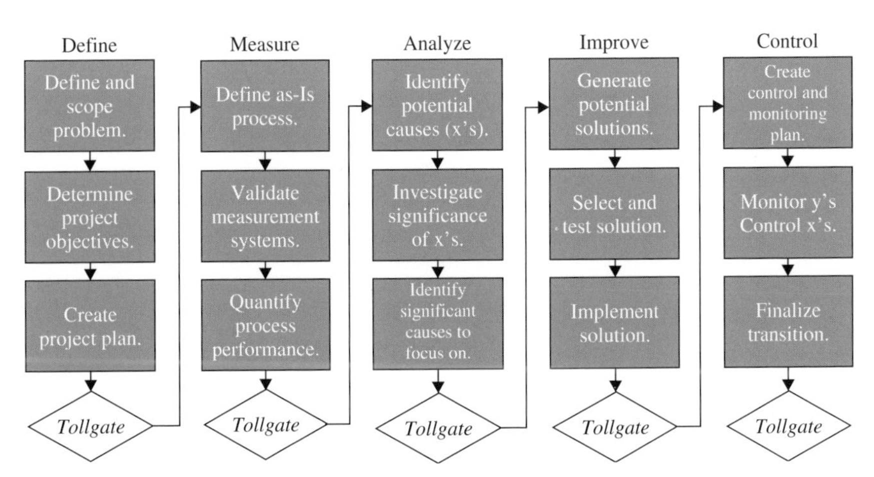

Figure 4.6 Integrated Lean Six Sigma Methodology.

We start with identifying the candidates for improvement in flawless execution. Then we move on to defining the value through the eyes of the customer. By creating *value stream maps* and *voice of the customer*, we can define the *customer critical to* (CTs) components. We then identify waste in our processes, which are process steps that don't add value for customer and have no value from environmental, safety, or regulatory perspective. These are eliminated or minimized.

We will also encounter many projects that require the rigor of probabilistic and statistical thinking. In every project, we follow the principle of "if you cannot measure, you cannot improve" philosophy. Also, we ask the question of if we have evidence that we can trust the data. This approach has been successfully applied in the public sector and in almost every industry spanning from retail to financial services to pharmaceutical to health care to high technology and electronics to nuclear power generation industries.

Chapter 5

Deploying Design for Lean Six Sigma

In this chapter, we will explore how we might create a system across the organization to deploy the principles of *Design for Lean Six Sigma* (DFLSS). If the principles of DFLSS were to become pervasive and the processes of DFLSS were to be followed rigorously, then we must create an infrastructure and establish a governance structure that promotes the deployment objectives.

5.1 DEPLOYING DFLSS

Every organization is involved in four types of key processes. They are associated with creating, improving, operating, and managing activities. While the traditional focus of Lean Six Sigma approach is on improvement activities, DFLSS is focused on creation processes. Therefore, it is important to link the DFLSS approach with company's innovation and design processes, resources, budget, and priorities.

Almost every company follows certain process for designing its products, services, business models, and processes. The first part of the deployment journey is to evaluate this process for its

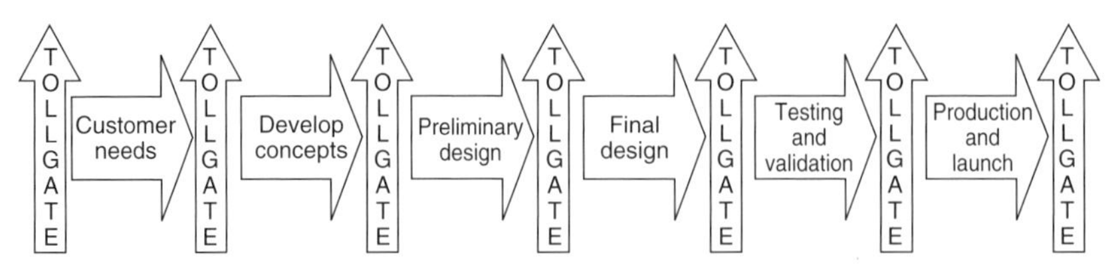

Figure 5.1 Typical Design Process with Phases and Tollgates.

current state. Many firms follow a traditional stage-gate process (also known as phase-gate process) in which the entire design process is divided into a discrete set of smaller phase. (See Figure 5.1.)

A typical design process consists of gathering customers' unmet needs, creating design concepts, developing the solution with design steps, testing and validating the design solution, and executing full production. Several variations of these steps exist and sometimes the design step is broken into preliminary design and final design, especially if an iterative approach is facilitated. Concept development, preliminary design, and final design are essentially the key steps of design activities.

Each of the phases is followed by a tollgate review to ensure that all key tasks associated with that phase are completed. Each phase has a well-defined set of objectives and tasks. The tollgate associated with each phase has a list of deliverables against each of the key activities.

For example, key activities associated with "gathering customer needs" phase typically include identification of customer and market requirements, translating customer wants into measurable critical-to customer expectations, review of technology capability, analysis of competition, establishment of possible configurations to meet customer expectations, review of the overall market opportunity, translating critical-to elements into functional requirements, establishing business models, and formalizing the team.

At the end of the customer needs phase, there is a tollgate review for evaluating the deliverables associated with each of the key activities. For example, some of the key deliverables for this tollgate includes a high level business case, prioritized list of customer needs, technology assessment report, preliminary project plan, functional requirements, competitive benchmark, documentation regarding potential solutions, and risk assessment report. The review will ensure that deliverables are completed on time and bear expected quality.

Design for Lean Six Sigma is not a replacement for the company's current design process. Instead, it should enhance the quality of the design process and reduce the development cycle times. Neither should it be treated as a set of stand-alone tools but as a system to be integrated with the existing design process. Toward this end, we must merge DFLSS activities and deliverables with the company's new product or service design process. Integration of DFLSS activities and deliverables with business or product specific activities and deliverables ensure that design process is optimized to produce the world-class results (see Figure 5.2).

Regardless of specific design activities and deliverables in each phase, we must consider the following factors for all gates review:

- Business case validation with financial analysis
- Team and resource readiness
- Project, financial and market risks
- Customer expectations regarding the product or service
- Management of knowledge and lessons learned
- Documentation/materials associated with design, regulatory, and customer activities

The decisions and questions associated with each phase of the integrated design processes are usually similar. We will use the

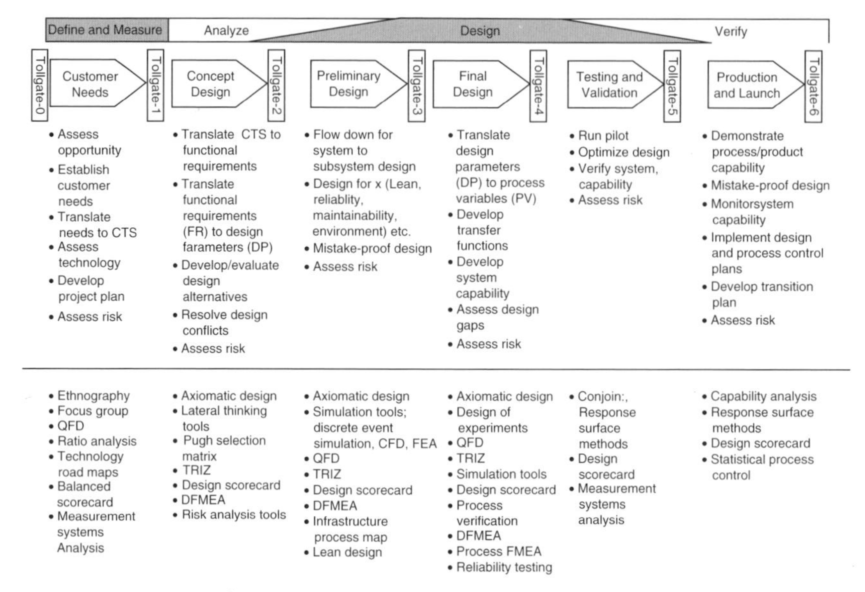

Figure 5.2 Sample DFLSS Activities and Tools Linked to Design Phases.

six-phase example discussed in this chapter to explain the key decisions and questions. (See Figure 5.3.)

The key decision for gate 0 is to decide whether to proceed, reject, or rework the case to initiate the new product or service development process for the proposed project. The key deliverables in this regard are high-level business case, customer needs and potential offering. Consider these key questions:

- What is the potential market opportunity, and why?
- Who are the customers for the proposed solution?
- Is the program sponsored by the business leadership?
- Does the proposed program align with business strategy?
- What is competitive environment in the marketplace?
- Is there a technology or business capability available for generating the solution?

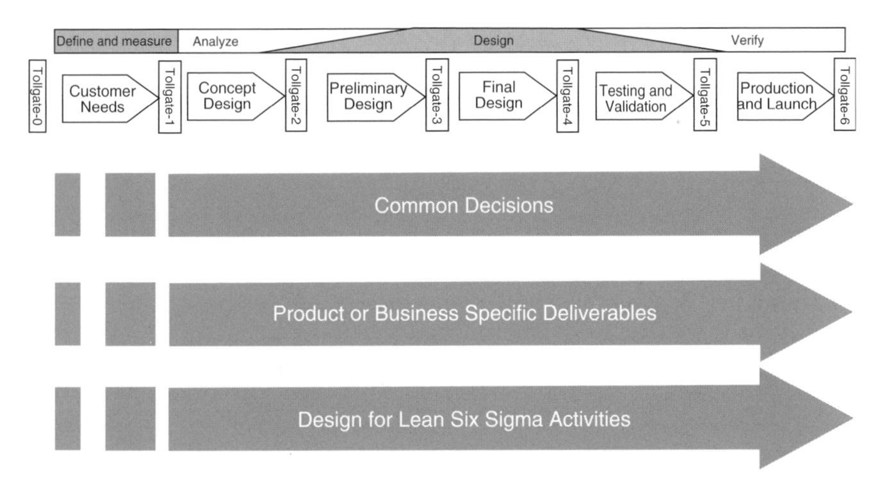

Figure 5.3 Integration of New Product Development Process with DFLSS Process.

The key decision for gate 1 (customer needs) is to decide if we have a valid business case, customer needs are well defined, and feasibility is likely. The key deliverables in this regard are detailed business case, customer needs, and technology or business capability assessment. Consider these key questions:

- What are the unmet customer needs (jobs to be done)?
- Who are the customers?
- Who are the stakeholders for the project?
- What is the strategic importance of this project?
- What is the size of business opportunity?
  - Market share
  - Time to market
  - Financial viability
  - Technology feasibility
- What are prioritized customer needs?
- What are business requirements?

The key decision for gate 2 (concept development) is to decide if we have a technically feasible solution that meets customer and business expectations, and necessary resources (financial, human, and technical) are available. The key deliverables in this regard are details of conceptual solution, functional requirements, competitive benchmark, and preliminary project plan. Consider these key questions:

- What key functions will the solution address?
- What are the most suitable conceptual designs?
- Are the concepts technically feasible?
- How well will the solution perform against customer expectations?
- What is the business impact?
- What data are available, and how trustworthy are the data collected?
- What are potential design options to support the concepts?

The key decision for gate 3 (preliminary design) is to decide if we have a preliminary design that is technically sound and feasible, a project plan that is viable, and the technical, financial, and customer risks that are still within acceptable limits. The key deliverables in support of the decision-making process are design scorecards, project risk assessment reports, system and subsystem design architecture, design matrix (functional requirements to design parameter), design of the supply chain configurations, and technology assessments. Consider these key questions:

- What is the expected performance of the design?
- What is the design of the system and subsystem architecture?
- How well do the design parameters address the functional requirements?

- Has a detailed integrated program/project plan been developed?
- What are the results of the Design Failure Modes & Effects Analysis (DFMEA)?
- What are the results from model/simulation/rapid prototypes?
- What are risks based on the design reviews?

The key decision for gate 4 (final design) is to decide if the stakeholders accept that we have a detailed design that is technically sound and feasible, and that meets customer, business, regulatory, and environmental requirements. The key deliverables in support of the decision-making process are design scorecards, updated project risk assessment reports, final design matrix (functional requirements to design parameters to process variables), design FMEA, test plans, transfer functions, configuration management, and capability flow up. Consider these key questions:

- What is the predicted performance of the product or service?
- What is the quality of the design based on the FR-DP mapping?
- Are all the necessary transfer functions established?
- Is the design robust to environmental degradation?
- What are the risks, based on the design reviews?
- Will the design satisfy customer expectations?
- What are the plans for piloting or prototyping?

The key decision for gate 5 (testing and validation) is to decide if the stakeholders agree that product, process, and/or service requirements have been demonstrated. The key deliverables in support of the decision-making process are pilot and prototype test results, design optimizations, and process capability information. Consider these key questions:

- What are the results of prototype or pilot?
- Have the objectives of the design been demonstrated and validated?
- Are there further opportunities for optimization based on the tests or prototypes?
- Is the pilot/prototype meeting customer and stakeholder expectations?
- Have all business, functional, and service concerns been addressed?
- Are validation plans ready?
- Is the supply chain ready and capable?

The key decision for gate 6 (production and launch) is to decide if the development process is completed and transition ownership is established. The key deliverables in support of the decision-making process are transition plans. Consider these key questions:

- What are the obstacles for transitioning to production and launch?
- Is the transition plan complete?
- Has the plan been communicated to the production team?
- Has functional support been established?

5.2 DESIGN FOR LEAN SIX SIGMA ENTERPRISE

Deployment of DFLSS is enabled by the supporting organizational infrastructure. Following are the key roles typically found

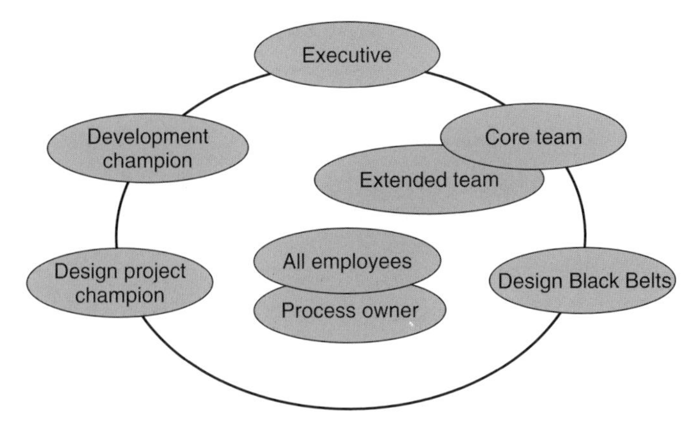

Figure 5.4 Roles within a Typical DFLSS Deployment.

inside a DFLSS enterprise, shown in Figure 5.4:

- Executive sponsor
- Deployment champion
- Design project champions
- Design Black Belt
- Core team
- Extended team

5.2.1 Executive Sponsors

The executive sponsor has the vision for achieving growth objectives through design and innovation. This person provides the direction to integrate DFLSS with the design process and lead the change efforts. It is the responsibility of this individual to remove hurdles and road blocks that might come against DFLSS deployment. The executive sponsor must provide guidance and counseling to the deployment champion. The sponsor must walk the talk by reviewing DFLSS project results and progress.

5.2.2 Deployment Champions

The DFLSS deployment champion provides day-to-day management and direction of DFLSS program. It is the responsibility of this individual to work closely with design process owners and develop the deployment strategy for DFLSS.

5.2.3 Design Project Champions

The main responsibility of design project champion is to identify and scope DFLSS projects. Depending on the size and scope, a new product or service to be introduced could be scoped as a single design project or multiple design projects.

5.2.4 Design Black Belts

Design Black Belts lead design projects through the application of DFLSS principles. In this capacity, they work closely with the core team responsible for the design projects as well as the extended team. Successful Design Black Belts will possess project management skills, leadership skills, knowledge, and experience of applying DFLSS principles and change management skills.

5.2.5 Core Team

Core team consists of a team leader, subject matter experts, and others who will work directly on the design project. It is this team who will be normally responsible for the execution of the project. Usually there is a core team with responsibility for the system or subsystem. To execute the project with DFLSS rigor members of this team should have Design Green Belt level of knowledge. Design Black Belts serve as mentors to this team.

5.2.6 Extended Team

This team supports the activities of the core team. Their primary objective is to support the core team and provide expertise to fulfill a specialized need. Depending on the nature and extent of the work they perform, this team may need to learn DFLSS at Design Green Belt level.

Life of Design Black Belts

Since the Design Black Belts serve a key role in the implementation of DFLSS-based projects, it is worth discussing some of their success factors as well as their major responsibilities. First of all, the core depends on the knowledge and skill of Design Black Belt on matters concerning technical aspect of DFLSS. They should be familiar with the design process and must demonstrate leadership in promoting innovation and project execution. They must possess bias for action in order to facilitate on-time project execution. They act as agents of change, working with champions and process owners as well as reporting out on progress to business leadership. In this sense, ability to influence without authority is a key desirable characteristic to possess. They are also adept at working with customers, suppliers, and process experts so that they can facilitate information flow-up and flow-down for design. In addition, they have great communication, teaching, and coaching skills to transfer knowledge to the core and extended team. They also assist champions and process owners in identifying project opportunities.

5.2.7 Building Support Infrastructure

A successful enterprisewide deployment requires the support of an infrastructure to manage, scale up, and sustain the DFLSS activities. Major infrastructure dimensions include program management, finance, human resource, communications,

IT, and training. There are policies, practices, and decisions to be made on each of these elements.

Program Management

The function of the program management is to develop and manage the portfolio of projects and programs deployed using DFLSS approach. Key decisions include how the pipeline is identified, deployed, funded, resourced, and executed.

Finance

Several key decisions are to be made on financial assessments and validation of DFLSS projects. How are financial benefits estimated for redesign projects and new design projects? How will we establish hard savings resulting from a redesign versus potential new growth in revenue from new projects? How will we account for cost avoidance from better designs?

Information Technology

IT plays a key role in DFLSS implementation with tools for automation, data management, and program management. Its policies can impact data collection, data access, and data management. In addition, there are several software packages that are needed for simulation activities as well.

Human Resources

HR must help with policies regarding recruiting of Design Black Belts and champions. A key to successful DFLSS deployment is the implementation of employee performance management. Performance objectives must be aligned with DFLSS plans. Other decisions include repatriation policies for Design Black Belts and champions, rewards, and recognition approaches for the team and DFLSS stakeholders.

Communications

How will we communicate the DFLSS plans, success stories, and issues internally? How will we promote the new competency and culture? What part of the messages can be communicated outside the organization? Will the savings and growth goal attached to DFLSS be communicated to external stakeholders?

Training

Who will manage the training needs and logistics for design black belt, design green belt, champion, and executive education? Who will address the curriculum design and material development? What are the plans for training the trainer and coaching the teams? These are some of the sample issues to deal with.

What is important is that we think through various policies and procedures and make decisions that are best suited for the company environment and culture. The best practice from another company may not be the best option for your company's environment.

Chapter 6

Capturing the Voice of the Customer

Businesses exist to create value for their stakeholders. For most corporations, their primary objective translates into creating wealth for its shareholders. Shareholders want profitability and growth in return for their investments. Therefore, companies have specific and tangible financial objectives for the short and long term. These financial objectives are achieved through creating unique value for the customers it serves (see Figure 6.1). The core of the company's strategy formulation is to establish the unique and differentiating value proposition it has for its customers. What is the service or product we will sell to customers that will satisfy their certain unmet needs better than any of our competitors? Who are these customers and what are their stated and unstated needs? Do we have a unique solution or capabilities that will address this gap in market place? Simply put, what's the promise we make to our customers?

Once we have established the value proposition for the customers, we must ask the question of what critical processes we must excel in to deliver flawlessly on the promise we made to the customer. This requires identifying, monitoring, managing, and improving essential processes so that we can excel at delivering

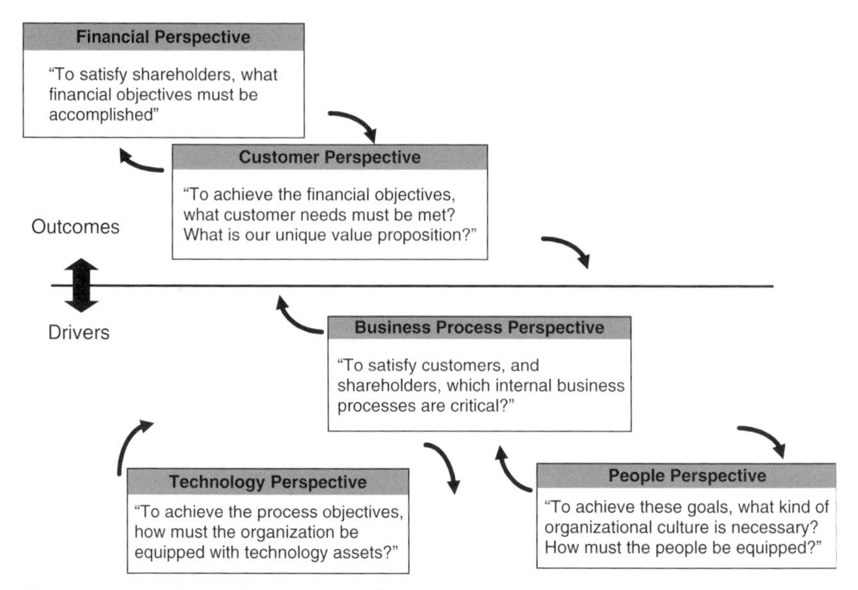

Figure 6.1 Organizational Value-creation Hierarchies.

on the promise. This, in essence, is the function of the Lean Six Sigma approach. Critical processes are identified, documented, measured, aligned, and improved to achieve the objectives. The alignment process involves creating ownership and accountabilities at organizational level. The improvement process involves identification, prioritization, scoping, and resourcing of the suitable candidates for improvement.

Technology and people are enablers to the business processes. The objective is to optimize the use of these resources for the efficient, speedy, and economic execution of processes. Therefore, human capital and information/technology capital must be equipped and managed to achieve maximum return for the organization.

Keeping this framework in mind, let's take a closer look at how organizations create value for their customers. Every organization makes a promise to its customers. As a result, customers have formed certain expectations regarding that promise. However, when the promise is delivered in the form of products and services, it is often strikingly different from what customers were expecting. The question is, then, how do we create the best experience for the customers that meets or exceeds their expectation through the delivery of the promise? To do this, we must understand customers' expectations and act on them through the delivery of superior products and services.

6.1 DEFINING ELEMENTS OF CUSTOMER–PRODUCER RELATIONSHIP

In the literature there is much confusion around the topic relating to customer and producer relationships. A good place to start would be to provide clarity around the labels for various concepts attached to this relationship. For example, there seems to be no standard operating definitions around the concept of *customer expectations*. Common labels attached to customer expectations are customer wants, needs, requirements, specifications, standards, demands, wishes, delighters, expectations, CTQ (critical to quality), and CTS (critical to satisfaction). Similarly, we use many names to address our customers. Based on what we do for them or with them and their roles, we call them names such as internal customer, external customer, client, patient, consumer, guest, constituent, fan, partner, patron, subscriber, dealer, or stakeholder. Deliverables provided to customers are often called products, services, service product, service process, goods, package, delivery, or outputs.

As a first step toward understanding the customer–producer relationship, let us define some key elements involved. They are illustrated in Figure 6.2. The most important ingredient in this relationship is the deliverables to the customer. For simplicity let's call it product. *Product* is something created or provided

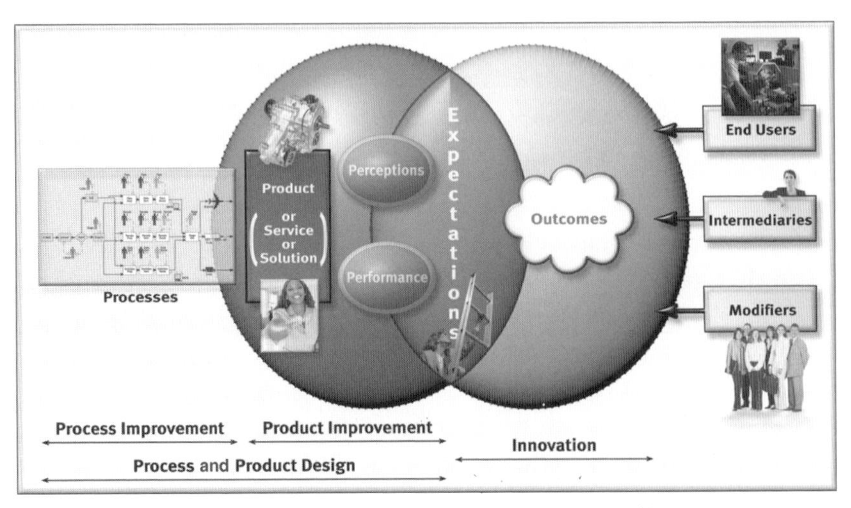

Figure 6.2 Producer–Customer Value Creation and Improvement Process.

by the producer for the customer. Customers use it to achieve certain results. It may be tangible or intangible. Therefore, all the intangible deliverables to customers in service industry are accommodated by this definition. It should be countable items and is typically made plural by adding the letter *s* at the end. For example *an answer* is a product, and the plural form of the product is *answers*. Other examples include drawings, shipments, diagnoses, reports, and paintings. Products are further classified into target products and final products. *Target products* are the products are the ones we have created for a certain customer(s) in mind. These customers may add functions or features to the product and transform them for use of other customers. These are called *final products*.

The next step is to define *customer*. Although there are many ways to name and define the customer based on power, influence, whether they are internal or external to the organization, what they receive from the producer, and so on, we use a simpler definition based on the role they play with reference to the product.

In this approach, there are three kinds of customers. They are end users, brokers or intermediaries, fixers or modifiers.

End users are the ones for whom the product was primarily created. They use the product to achieve certain desired objectives. For example, students are end users of lectures created and delivered by the professor. Other examples include hospital patients, hotel guests, taxpayers, and sports fans.

Brokers or intermediaries transfer the product from producer to end users. They are the middle-person(s) in the relationship between end users and producers. As an agent of the producers, brokers make the product accessible to end users and encourage the proper use of it. Brokers transmit end users' expectations to producers, make the product easy to access, and enable the transaction between the producer and end users. Pharmacists, travel agents, automobile dealerships, patient advocates, and bookstores are examples of this role.

A *fixer or modifier* repairs, corrects, or modifies the product in order to meet the expectations of the end users or brokers. This action can happen anytime during the life cycle of the product. Examples of this role include call center representatives, service technicians, pharmacists, and repair centers. The same person may play many of these three roles.

Producer is the one who creates or acquires the product for delivering to the customers. The producer makes a direct or indirect promise to the customer. The promise is fulfilled through the delivery of the product to the customer. *Process* is the mechanism by which inputs are transformed to outputs. Since products are one kind of outputs, it implies that process is the mechanism by which the products are created. Auto manufacturers are producers of cars, and various operations such as welding, milling, materials acquisitions, and assembling are examples of processes.

A key step toward flawlessly delivering on the promise to the customer is to understand various elements within the producer–customer relationship. It starts with defining the key product or product categories. For example, *key products* involved in the context of physician–patient relationship are diagnoses, advice, prescriptions, reports, invoices or bills, and surgeries. Let's take the example of prescription as a target product. It is the piece of paper that has information about the medication for the patient. The physician is the *key producer* of this product. The end user for the product is a certain pharmacist, since it intended for his or her use. The pharmacist uses it as an input to assemble another specific product called medication for the patient. The patient is one of the brokers in this transaction, since the patient is completing the transaction of delivering the prescription to the pharmacist. The nurse, physician, or pharmacist may act as the fixer by correcting any errors in the prescription or clarifying the information for the end user, namely the pharmacist.

The promise to the customer is fulfilled through the delivery of the product to the customer. The customer has formed certain expectations regarding the product. Customer satisfaction is the degree to which these expectations are met. It turns out that there are two kinds of expectations customers have formed regarding the product. Producers focused in satisfying the customers proactively seek out and gather these expectations in order to improve the characteristics of the product.

6.2 CUSTOMER EXPECTATIONS

Customer expectations and the degree to which they are satisfied form the basis of customer satisfaction. There are three types of customer expectations. They are performance, perception, and outcome expectations. The first two types relate to the products delivered to the customer, and hence are also called *product expectations*. The outcome expectations exist in a solution-neutral environment. These expectations exist even in the absence of products.

Let's explore them in greater detail. *Performance expectations* are objective, unambiguous, and measurable expectations that customers have regarding the product. Examples of this include weight of the product, product delivery lead times, cost of the product, and product quality characteristics such as reliability measured in mean time between failures. The other product expectation, called *perception expectation*, is composed of the subjective, ambiguous, and difficult-to-measure characteristics. Examples of this type of expectations include ease of use, look and feel, timeliness, ease of doing business, and durability. Many of these expectations become performance expectations, once we establish operational definitions and associated measurements. The third type of expectation, called *outcome expectation*, is the result customers want to achieve by using the product or working with the provider. Examples of this type include health, fun, market share, wealth, and return on investment. It is important to note that these expectations – fun, entertainment, health, and others – are solution neutral and exist even in the absence of the products. These expectations are regarding the *jobs to be done*, as the name implies (Christensen et al. 2007).

By identifying, measuring, and acting on the performance and perception expectations customers have regarding the product, we can *improve the quality of the product*. This, in essence, is the key step toward creating a flawless delivery of the promise. However, through identifying and measuring customers' unmet outcome expectations and exploring better solutions that satisfy these outcomes, one can *accelerate innovation* and thus create new market spaces. Both sets of activities are essential for advancing the agenda of the organization by delivering flawlessly on the promise and continuously creating new and better promises.

Let's look at another example of how we might apply this model. Let us assume that our company sells lawn mowers. Our company wants to be customer centric and decides to proactively gather customer expectations. We start with identifying the customers of a particular model of lawn mower we sell. These customers are categorized into end users, intermediaries (also called brokers), and modifiers (also called fixers). Then we will solicit the expectations from each of these customer groups regarding the product as representing various stakeholder objectives. Customer expectations can be collected through multiple mechanisms; one of the effective approaches is the focus group method.

Table 6.1 indicates a partial summary of product expectations from the end users collected using the focus group method. Some of these expectations are perception expectations that are ambiguous and hard to measure, while others are performance expectations that are objective and unambiguous. By capturing these expectations, one can design or improve the quality of the lawn mower, as in this example. In the forthcoming chapters we will explain how we can use these customer expectations to design new products or solutions by converting them into functional requirements and later into design parameters.

In a similar way, we can explore the outcome expectations regarding the lawn mower. Outcomes are the results customers want to achieve by using the product or working with the provider. There are two types of outcome expectations – desired outcomes and undesired outcomes. Desired outcomes are the ones customers want to achieve and the undesired outcomes are the ones customers want to avoid. Similarly, there are outcome expectations from the provider – desired and undesired ones.

Table 6.2 provides an example of desired and undesired expectations from customers and providers. It is important to note that these expectations exist in solution-neutral environments.

Table 6.1 Sample End User Expectations of a Lawn Mower

Expectations	*Type of Expectation*	*Measure*	*Units*	*Target*	*Current Level*	*Gap*
Energy efficient	Performance	Fuel efficiency	MPG			
Easy to push	Perception	Pushing force	Lbf			
Noise	Performance		db			
Weight	Performance		Lb			
Easy to store	Perception	Volume	Cu. Ft			
Cuts grass at different heights	Perception	Height adjustment options	#			
Reliable	Performance	MTBF	Hrs			
Low emission	Performance	Nox, CO, and UHC Levels	PPM			
Easy to maintain	Perception	# of repairs	#			
Ease of access for repair	Perception	Component modularity	Rating (1 to 10)			
Appealing color	Perception	Rating	Rating (1 to 10)			

These needs exist regardless of whether lawn mowers exist or not. Knowing the outcome expectations are the keys to successful innovation. In this example, one can anchor at the desired outcome of creating a beautiful looking lawn that will maintain its height at a certain fixed level. The outcomes to be achieved are also called the jobs to be done. In the context of innovation, our objective should be to identify these jobs to be done in unoccupied market space. Then the next step would be to explore possible solution spaces. There are many approaches to explore the new paradigm or solution. One such solution for this

Table 6.2 Sample Desired and Undesired Outcome Expectations from Customers and Providers

	Desired Outcomes	*Undesired Outcomes*
Customer	Consistently green-looking lawns	Utilize too much energy to maintain the grass
	Keeps certain grass height throughout	Uses too much water
	Grass looks attractive	Noise pollution
	Easy to maintain	Environmental pollution
	Smells great	Pollution from end-of-life discard
	Robust to harsh conditions or usage	Allergenic
		Costs too much to own

	Desired Outcomes	*Undesired Outcomes*
Provider	Consistent revenue growth	Product liability/ lawsuits
	Predictable profit	Imitation products
	Customer loyalty	Environmental complaints
	Steady demand	Supply shortages
	New derived products	

example is genetically engineered grass seed. Is this solution superior to a lawn mower? What is the measurement by which we will assess an innovation?

Although there are many measures, one convenient and superior measure is the concept of ideality. *Ideality* is defined as the ratio of desired outcomes to undesired outcomes. Our objective is to increase the value of ideality through innovation. We achieve this by improving the benefits (desired outcomes) and/or reducing the cost and harm (undesired outcomes). In the previous example, genetically engineered grass seed has the potential of improving the benefits and reducing the cost and harm; hence arguably a better innovation.

Let's look at another example. Prior to the release of Quicken software by Intuit Corporation, customers had the option of using expensive accounting software or manually perform the calculations using paper and pencil for managing personal finances. On one hand, pencil and paper provided two important desired outcomes – low cost of ownership and ease of use. But pencil and paper alone failed to provide other desired outcomes, such as accuracy and reliability of transactions, speed and flexibility with financial calculations, and the capability to perform analysis about personal finances. On the other hand, accounting software provided better solution for desired outcomes not achieved by pencil and paper. But it was expensive and hard to use because of specialized accounting jargon. By gaining insights into the details of the job to be done, Intuit was able to create a newer solution that increased the ideality of the innovation. By combining the desired outcomes from the substitute products, Intuit was able to offer the Quicken product and so create a new set of values for their customers.

6.3 METHODS OF COLLECTING CUSTOMER EXPECTATIONS

Common methods for collecting customer expectations in a proactive way are surveys, interviews, focus groups, and observations. In addition, we can use existing information such as complaints or feedback from customers that might be available from call center data, warranty and product return data, customer service representatives, and sales representatives. It is also important to know the different customer voices available from end users, brokers, and fixers. For example, sales representatives, resellers, and third-party and industry experts might play the role of brokers. Other groups such as customer service representatives, call center associates, governmental and regulatory agencies, companies' internal departments, and industry experts might play the role of fixers. Seeking information from all these sources enables us to create better value for the customer or improve the value propositions for our existing offerings.

In most situations, we start customer research with some information already regarding customer expectations. We continue gathering more information using many different methods. Depending on the situation or the stage of data collection, certain method is more appropriate than another. For example, interview method is usually used during early phases of learning about a particular customer segment. This method is very effective when we discover new customer segments and do not have a hypothesis as to their needs. On the other hand, the focus group method is very useful to gather a collective point of view from several customers at the same time. Usually we have certain high-level hypotheses concerning their expectations while utilizing the focus group method. Survey approach is typically used to measure customer priorities on a scale large enough to draw statistically valid information to base business decisions on. Surveys are most useful when they have already developed specific hypotheses about customer expectations using other means such as focus group method. Observations are most powerful in order to gather unarticulated customer expectations. Therefore, observations are very important for proactively collecting customer outcomes. A powerful way of conducting observations to gather customer expectations is by utilizing ethnographic research. *Ethnographic research* has its roots in cultural anthropology. The research focuses on the sociology of meaning through close field observations of sociocultural phenomena.

Interview method is used at the beginning, transitioning, or end phase of customer research. At the beginning phase, it is used to learn about varying expectations customers have and to develop certain potential hypotheses regarding their expectations. During the middle or transitioning phase of customer research, interviews can clarify why a certain issue is particularly important to customers. This approach is very useful during the latter part of customer research in clarifying certain findings, seeking new ideas or suggestions, or piloting a solution and collecting feedback. The most important point to keep in mind is that the interview method is best suited for clarification and discovery. It is usually not used to make business decisions yet. Think of the interview as more a guidance system to help shape a focus group or survey.

A focused approach of interview techniques provides flexibility and is excellent for seeking clarifications. It can be conducted telephonically or in person. We can expect high response rate. However, it can be time consuming and costly to execute, resulting in smaller samples. The biggest drawback of the interview approach is the influence of interviewer bias. For example, loaded, ambiguous, or leading questions can result in erroneous conclusions or decisions. Other issues include overspecificity and overgeneralization, utilization of nonrandom samples, and sequencing of questions. Our objective should be to minimize the noise effects of these bias factors.

Focus group method is one of the best methods for gathering customer needs. It is a powerful mechanism to gain insights into the prioritization of customer expectations. It is typically used as a next step after conducting customer interviews in order to develop certain hypothesis or as a preliminary step in a survey process to gather quantitative information regarding customer expectations. Also, this method is often used to test concepts and get feedback from a pilot study.

A focus group is assembled to represent a customer segment with similar needs. The segmentation has less to do with gender, socioeconomic strata, or other demographic representation. A customer segment is a group of people with similar needs regarding a job to be done. An example of a job to be done is the organization and management of people's personal finances. People want to manage their expenses such as credit card expenses, auto, mortgages and other loans, as well as investments in savings, checking, and money market accounts, retirement funds and stocks or bond markets. This is different from another job to be done in the corporate context, namely managing corporate finances. Customers who have the need of managing personal finances share many common expectations. They should be treated as same segment, regardless of gender, socioeconomic status, or other demographics.

A typical focus group session consists of seven to twelve participants who share the needs of a job to be done. It will usually last one to four hours. The session is often repeated at least three or four times with different groups to avoid sampling errors. The most common objective of the focus group is to gather feedback from customers regarding product attributes, product expectations (performance or perception expectations) or outcome expectations (jobs to be done). Participants' feedback is requested in greater depth regarding a well-defined area of interest. Three key phases of the focus group are planning, conducting, and analyzing the focus group session. The key activities during the planning phase are deciding on the objectives of the session, establishing the participants and other resources needed, creating the agenda and flow, selecting locations and schedules, and developing and finalizing the questions.

The format of the focus group depends on the objective of the session. If the objective is to get customers' opinions regarding product expectations, then the major emphasis is to describe the solution concept and then seek customers' expectations regarding the product. These are typically performance expectations such as weight, cycle time, cost, and other quality characteristics or perception expectations such as easy to use, appealing, and great value. If the objective of the session is to get the users' input directly on the design of the product or solution, then we would ask the participants to describe the attributes of a satisfying product or solution. However, if the objective is to gather the outcome expectations, then we would explore why they need the product or solution or what job they are trying to get done. By understanding what makes them successful, we have the necessary starting point for driving innovation.

Although the focus group method provides great flexibility and enable high response rate, it can be influenced by moderator bias and dominant personalities. Also, the sample size is typically small and the method suffers from issues relating to the design of questions. Watch for issues pertaining to loaded questions, leading questions, and question sequencing.

Survey method is very useful for gathering a considerable amount of information from a large population in order to draw statistically valid information on which to base business decisions. It is often used to measure as-is conditions, as well as changes and causality. There are several ways of conducting surveys; the main approaches are manual methods administered via mail, phone, or in person, and electronic methods administered via Internet, e-mail, or automated telephone methods. The best approach depends on various factors such as cost, interviewer bias, anonymity, time available to collect data, and the need to obtain open-ended responses. Surveys can usually be administered with relatively low cost, and quantitative data can be collected with large sample sizes. Although the response rate is typically low (20 to 30 percent is common), it minimizes interviewer or moderator bias. The number of questions in the survey is often limited in order to gain a higher response rate.

The ethnographic research method of collecting customer expectations relies heavily on field work conducted through observations, personal experiences, and participation with customers. As mentioned earlier, ethnography has its roots in anthropology and sociology. It assumes that human behaviors and expectations can be understood by studying the culture or small group they belong to. Ethnographic researchers are trained to use a wide range of qualitative and quantitative data collection. Usually, the method uses three kinds of data collection in this context – observations, interviews, and documents. A narrative description is typically created containing quotes, descriptions, and excerpts. The graphs, charts, and various artifacts together form the story of the culture or subculture.

More and more companies are using this approach of gathering customer expectations owing to the success with collecting unarticulated needs. For example, one hospital group uses ethnographic research methods to understand its customers and thereby create better designs for its hospitals and newer solution offerings. The researchers spend time looking at various subcultures of people coming to the hospital. When an ethnographer camps out in the hospital room with an elderly cancer patient undergoing a surgery, he or she can gain valuable insights through observation and interaction. Through interaction and observation of human behavior, ethnographic researchers develop deeper understanding of how we interact with spaces, equipment, people, and surroundings. This, in turn, leads to new learning, which translates to better health care services.

The focus of ethnographic research is typically conducted with a small group so that people's behavior can be studied carefully in everyday contexts, rather than under experimental conditions created by the researcher. Although data collection involves observations, interviews, and documents or artifacts gathering, the approach is fairly unstructured and open ended to enable flexibility and minimal interruption of natural settings. The skill and training of the ethnographer is essential to the analysis of data and interpretation of the meanings and functions of human actions.

6.4 RESEARCH ETHICS

Since collecting voice of the customer takes place through two-way interaction with real human beings, there are a number of ethical concerns that we must address. First of all, researchers or facilitators must disclose the intent of study or research and gain informed consent to the research beforehand. It is important to let the participants know how the data from the study will be used. It is also important to know whether the participants want to remain anonymous or may be named in the written report. The other key principle is about maintaining confidentiality. We must ask the question of who will see the results of the data. How long will they be kept? What are the implications of using the data other than what the research was intended? This is especially critical if you gathering data from employees in the company. Confidentiality and anonymity go hand in hand. Another point is about fulfilling any promise made to the participants. For example, the leader facilitating the survey or other instrument has an obligation to meet commitments on feedback, actions, and accountability. Decide in advance how this will be handled and who is responsible for it. In essence, facilitators must ensure that the research or study does not harm or exploit those among whom the study is done.

Chapter 7

Design Axioms and Their Usefulness in DFLSS

Axiomatic design (AD) theory (Suh, 2001) has been used in developing software, hardware, machines and other products and manufacturing systems. The axiomatic design theory has been used for four purposes:

1. To provide a systematic way of designing products and large systems
2. To reduce the random search process
3. To determine the best designs among those proposed
4. To create systems architecture that completely captures the construction of the system functions

Axiomatic design theory can also be used to map the customer domains and functional domains in the place of quality function deployment (QFD). Axiomatic design theory can be more effective in translating customer needs into functional requirements because we can study the relationships between functional requirements and design parameters with the help of design equations. A more elaborate discussion on axiomatic design is provided in the next section.

7.1 DESIGN AXIOMS

The theory of axiomatic design is based on two axioms:

1. Independence axiom
2. Information axiom

The *independence axiom* states that the independence of functional requirements (FRs) must always be maintained. FRs are defined as the minimum set of independent requirements that characterize the design goals (Suh, 2001). The second axiom states that the best design is the one with least information content and at the same time it should satisfy independence axiom. The information content is defined in terms of probability.

The information axiom can be best explained with the help of Figure 7.1.

The probability of success is calculated by using design range (usually tolerance) and system range (described by process

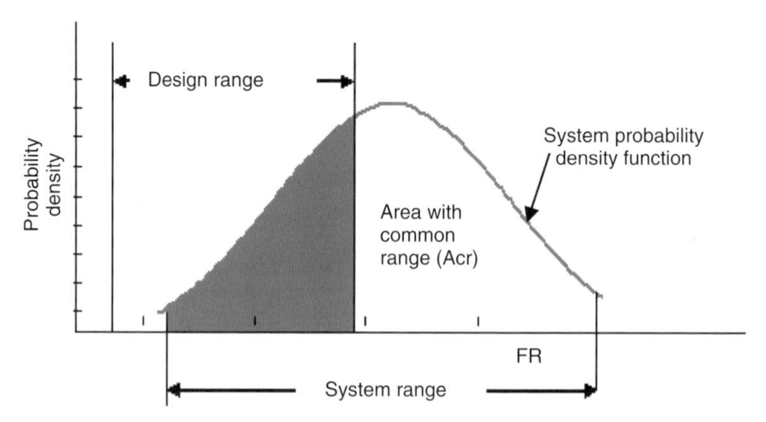

Figure 7.1 Information Axiom and Information Content.

variation), as shown in Figure 7.1. The information content is calculated by finding the area under common range (Acr) and is given by the following equation:

$$\text{Information content} = I = \log_2(1/Acr) \qquad (7.1)$$

From equation (7.1), it is clear that if the Acr = 1 or design range is equal to the system range, then information content is zero, indicating that the design is the best. With this argument we can say any design is good as long as system range is within design range, irrespective of process variation. This is the reason why knowledge about variation, Six Sigma, and DFLSS are important in the selection of best design with least variation.

7.2 DOMAIN THINKING

The axiomatic design world consists of four domains: the customer domain, the functional domain, the physical domain, and the process domain. Design is defined as interplay between what we want to achieve and how we achieve it. These four domains provide an important foundation for axiomatic design. The domain structure is shown in Figure 7.2. The domain on the

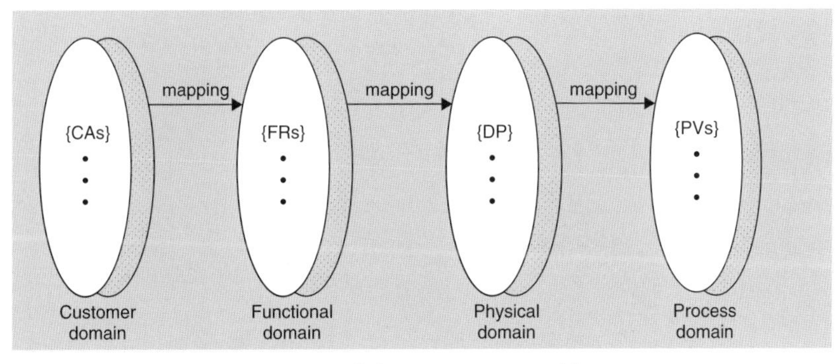

Figure 7.2 Four Domains of the Design World.

left relative to a particular domain represents what we want to achieve, whereas the domain on the right represents the design solution, how we achieve it.

In Figure 7.2, the bracketed abbreviations represent characteristic vectors of each domain. They are customer attributes {CAs}, functional requirements {FRs}, design parameters {DPs}, and process variables {PVs}.

In the customer domain we capture the needs (or attributes) that the customer is looking for in a product or systems. In the functional domain, the customer needs are translated to functional requirements (FRs). In order to satisfy the specified FRs, we identify design parameters (DPs) in the physical domain. Finally, to produce the product specified in terms of DPs, we develop a process that is characterized by process variables (PVs) in the process domain. Typically, the design equations are written in the following form:

$$\{\mathbf{FR}\} = [A]\{\mathbf{DP}\} \tag{7.2}$$

We can write down similar design equations for DPs and PVs. In equation (7.2), [A] is called the design matrix. If we have three FRs and three DPs, the design matrix will be as follows:

$$[A] = \begin{bmatrix} A11 & A12 & A13 \\ A21 & A22 & A23 \\ A31 & A32 & A33 \end{bmatrix}$$

The elements of the matrix [A] represent sensitivities, and they can be expressed using partial derivatives. For linear designs, Aij are constants. Therefore, if we assume linear relations between FRs and DPs, then Aij can be written as

$$\mathrm{Aij} = \partial\mathrm{FRi}/\partial\mathrm{DPji} \tag{7.3}$$

If we expand equation (7.2), we can write the following linear equations:

$$\text{FR1} = \text{A11 DP1} + \text{A12 DP2} + \text{A13 DP3}$$

$$\text{FR2} = \text{A21 DP1} + \text{A22 DP2} + \text{A23 DP3}$$

$$\text{FR3} = \text{A31 DP1} + \text{A32 DP2} + \text{A33 DP3}$$

Based on the structure of design matrix, we will have three types of designs. These designs are known as uncoupled designs, decoupled design, and coupled designs. Ideally, uncoupled designs are desirable because their design matrix will be a diagonal matrix indicating that every FR can be satisfied by one particular DP. These designs satisfy the requirements of independent axiom. Since it is not easy to come up with uncoupled designs, one may prefer to have decoupled designs. The decoupled design matrices are upper or lower triangular matrices. Decoupled designs will allow us to fix DPs in a particular order to satisfy the given FRs. By doing so, we can satisfy the requirements of independence axiom. All other design matrices indicate coupled designs. For a 3FR and 3DP case, the uncoupled and decoupled design matrices will have the following structure:

$$[A] = \begin{bmatrix} A11 & 0 & 0 \\ 0 & A22 & 0 \\ 0 & 0 & A33 \end{bmatrix}$$

Uncoupled Design

$$[A] = \begin{bmatrix} A11 & 0 & 0 \\ A21 & A22 & 0 \\ A31 & A32 & A33 \end{bmatrix}$$

Decoupled Design

In AD, we move from one domain to another domain in a zigzag fashion (as shown in Figure 7.3) so that we break down the requirements to the lower level. Zigzagging helps to decompose the functional and the physical domains and create the FR and

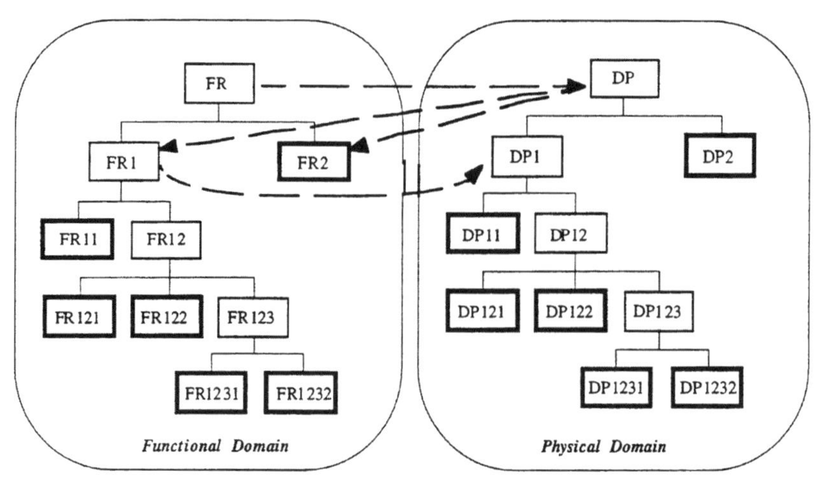

Figure 7.3 Zigzag Decomposition in Axiomatic Design.

DP hierarchies. The same method can be applied to decompose physical and the process domains and create the DP and PV hierarchies.

7.3 DESIGNING OF MTS SOFTWARE SYSTEM

Since software design is a good example of system design, we describe use of these techniques in system design by considering software design as an example. This particular software is intended to perform a multivariate data analysis called Mahalanobis–Taguchi Strategy (MTS). MTS is a pattern analysis tool that is useful to recognize and evaluate various patterns in multidimensional cases. Examples of multivariate systems are medical diagnosis systems, face/voice recognition systems, and inspection systems. Since in this technique Mahalanobis distance and robust design principles (Taguchi methods) are integrated, it is known as MTS. A detailed discussion of MTS is provided in Chapter 12.

Basically, there are four stages in MTS (Taguchi and Jugulum, 2002):

Stage I: Construction of a Measurement Scale

- Select a reference group with suitable variables and observations that are as uniform as possible. The reference group is also known as the Mahalanobis space (MS).
- Use the reference group as a base or reference point of the scale.

Stage II: Validation of the Measurement Scale

- Identify the conditions outside the reference group.
- Compute the Mahalanobis distances of these conditions and check if they match with the decision maker's judgment.
- Calculate signal-to-noise ratios (S/N ratios) to determine accuracy of the scale.

Stage III: Identify the Useful Variables (Developing Stage)

- Find out the useful set of variables using various combinations of variables with help from orthogonal arrays and S/N ratios.

Stage IV: Future Diagnosis with Useful Variables

- Monitor the conditions using the scale, which is developed with the help of the useful set of variables.
- Based on the values of Mahalanobis distances, take appropriate corrective actions.

In MTS, the Mahalanobis distance (MD) can be calculated by using the following equation:

$$MD = D^2 = (1/k)Z_i C^{-1} {Z_i}^T \tag{7.4}$$

where Z_i = standardized vector obtained by standardized values of X_i $(i = 1, \ldots, k)$

$Z_i = (X_i - m_i)/s_i$;

X_i = value of ith characteristic

m_i = mean of ith characteristic

s_i = standard deviation of ith characteristic

k = number of characteristics/variables

T = transpose of the vector

C = Correlation matrix

Based on these stages, the top-level FRs and DPs are identified:

FR = Perform MTS analysis
DP = MTS software program
FR1 = Construct measurement scale
FR2 = Validate the measurement scale
FR3 = Identify useful variables (screening)
FR4 = Conduct confirmation run (for future diagnosis)
DP1 = Reference group, variables
DP2 = Observations outside the reference group
DP3 = Screening method
DP4 = Useful variables

The corresponding design equation is:

$$\begin{bmatrix} FR1 \\ FR2 \\ FR3 \\ FR4 \end{bmatrix} = \begin{bmatrix} X & 0 & 0 & 0 \\ X & X & 0 & 0 \\ X & X & X & 0 \\ X & X & X & X \end{bmatrix} \begin{bmatrix} DP1 \\ DP2 \\ DP3 \\ DP4 \end{bmatrix}$$

From this design equation, it is clear that the design is a decoupled design. That means we can fix the order of the DPs to satisfy the FRs independently. In the decomposition of FR1-DP1, we can easily see that the design equations at lower level are also decoupled designs.

FR1-DP1 Decomposition

FR1—Construct scale; DP1: Ref. Group, variables

FR11 = Construct Mahalanobis space
FR12 = Obtain correlation structure
FR13 = Compute distances

DP11 = Variables, sample size
DP12 = Algorithm for correlations
DP13 = Algorithm for distances

$$\begin{bmatrix} FR11 \\ FR12 \\ FR13 \end{bmatrix} = \begin{bmatrix} X & 0 & 0 \\ X & X & 0 \\ X & X & X \end{bmatrix} \begin{bmatrix} DP11 \\ DP12 \\ DP13 \end{bmatrix}$$

→ **Decoupled Design**

PV12 = Equation for correlations
PV13 = Equation for distances

$$\begin{bmatrix} DP12 \\ DP13 \end{bmatrix} = \begin{bmatrix} X & 0 \\ X & X \end{bmatrix} \begin{bmatrix} PV12 \\ PV13 \end{bmatrix}$$

FR111 = Standardize variables
DP111 = Algorithm for standardization
PV111 = Equation for standardization

Similarly other top-level FR-DP decompositions are carried out. The details of these decompositions are shown here:

FR2-DP2 Decomposition

FR2—Validate scale; DP2: Observations outside ref. group

FR21 = Compute abnormal distances
FR22 = Compute S/N ratios

DP21 = Algorithm for MDs
DP22 = Method to get S/N ratios

$$\begin{bmatrix} FR21 \\ FR22 \end{bmatrix} = \begin{bmatrix} X & 0 \\ X & X \end{bmatrix} \begin{bmatrix} DP21 \\ DP22 \end{bmatrix}$$

PV21 = Equation for MD
PV22 = Equation for S/N ratio

$$\begin{bmatrix} DP21 \\ DP22 \end{bmatrix} = \begin{bmatrix} X & 0 \\ X & X \end{bmatrix} \begin{bmatrix} PV21 \\ PV22 \end{bmatrix}$$

FR3-DP3 Decomposition

FR3—Identify useful variables; DP3: Screening method

FR31 = Select suitable OA
FR32 = Perform MTS analysis

DP31 = Algorithm for OA selection
DP32 = MTS procedure for each run

$$\begin{bmatrix} FR31 \\ FR32 \end{bmatrix} = \begin{bmatrix} X & 0 \\ 0 & X \end{bmatrix} \begin{bmatrix} DP31 \\ DP32 \end{bmatrix}$$

FR32-DP32 Decomposition

FR321 = Generate MS
FR322 = Calculate distances
FR323 = Calculate S/N ratios
FR324 = Obtain avg. responses

DP321 = Variables, correlations
DP322 = Distance calculation method
DP323 = S/N ratio calculation method
DP324 = Avg. res. analysis

$$\begin{bmatrix} FR321 \\ FR322 \\ FR323 \\ FR324 \end{bmatrix} = \begin{bmatrix} X & 0 & 0 & 0 \\ X & X & 0 & 0 \\ 0 & X & X & 0 \\ 0 & 0 & X & X \end{bmatrix} \begin{bmatrix} DP321 \\ DP322 \\ DP323 \\ DP324 \end{bmatrix}$$

FR4-DP4 Decomposition

FR4—Conduct confirmation run; DP4: Useful variables

FR41 = Identify useful variables
FR42 = Generate MS
FR42 = Calculate distances

DP41 = Method of selecting useful variables
DP42 = Variables, correlation structure
DP43 = Method of calculating distances

$$\begin{bmatrix} FR41 \\ FR42 \\ FR43 \end{bmatrix} = \begin{bmatrix} X & 0 & 0 \\ X & X & 0 \\ X & X & X \end{bmatrix} \begin{bmatrix} DP41 \\ DP42 \\ DP43 \end{bmatrix}$$

We can easily see that at all levels of requirements flow-down process, we have either decoupled or uncoupled matrices, thus satisfying the requirements of independence.

The modules M1, M2, M3, and M4 of this design architecture can be represented as shown in Figure 7.4. These four modules correspond to construction, validation, screening, and confirmation respectively.

$$\begin{bmatrix} FR1 \\ FR2 \\ FR3 \\ FR4 \end{bmatrix} = \begin{bmatrix} A11 & 0 & 0 & 0 \\ A21 & A22 & 0 & 0 \\ A31 & A32 & A33 & 0 \\ A41 & A42 & A43 & A44 \end{bmatrix} \begin{bmatrix} DP1 \\ DP2 \\ DP3 \\ DP4 \end{bmatrix}$$

$$FR1 = M1DP1 \qquad M1 = A11 \longrightarrow \text{Construction}$$

$$FR2 = M2DP2 \qquad M2 = A21\frac{DP1}{DP2} + A22 \longrightarrow \text{Validation}$$

$$FR3 = M3DP3 \qquad M3 = A31\frac{DP1}{DP3} + A32\frac{DP2}{DP3} + A33 \longrightarrow \text{Screening}$$

$$FR4 = M4DP4 \qquad M4 = A41\frac{DP1}{DP4} + A42\frac{DP2}{DP4} + A43\frac{DP3}{DP4} + A44 \longrightarrow \text{Confirmation}$$

Figure 7.4 Modules in MTS Software Design Architecture.

7.4 DESIGNING A SYSTEM THAT WILL MARKET SPORTING GOODS

Design a system for a new Internet company that plans to market sporting goods (such as golf clubs and tennis rackets) over the Worldwide Web.

This is one solution to the problem. There can be many solutions, depending on FRs identified.

FR = Market sporting goods
DP = Internet-based company

FR1 = Identify market segments.
FR2 = Provide means to view products online.
FR3 = Provide easy and secure access for customers.
FR4 = Deliver products to customers on time.
FR5 = Provide online assistance to the customers.
DP1 = Market research
DP2 = Web site design
DP3 = User-friendly and secure transaction system
DP4 = Product delivery system
DP5 = Online help

At this level, the design equation can be written in the form of a decoupled design:

$$\begin{bmatrix} FR1 \\ FR2 \\ FR3 \\ FR4 \\ FR5 \end{bmatrix} = \begin{bmatrix} X & 0 & 0 & 0 & 0 \\ X & X & 0 & 0 & 0 \\ 0 & X & X & 0 & 0 \\ 0 & X & X & X & 0 \\ X & X & 0 & X & X \end{bmatrix} \begin{bmatrix} DP1 \\ DP2 \\ DP3 \\ DP4 \\ DP5 \end{bmatrix}$$

FR2-DP2 decomposition

FR21 = Enable important goods appear on the cover page.
FR22 = Enable customers to search for the product.
FR23 = Provide descriptions of alternatives.

DP21 = Cover page design
DP22 = Search algorithm
DP23 = Algorithm to display alternatives based on customer inputs

The corresponding design equation can be written as follows:

$$\begin{bmatrix} FR21 \\ FR22 \\ FR23 \end{bmatrix} = \begin{bmatrix} X & 0 & 0 \\ X & X & 0 \\ 0 & X & X \end{bmatrix} \begin{bmatrix} DP21 \\ DP22 \\ DP23 \end{bmatrix}$$

FR22-DP22 decomposition

FR221 = Enable customers to search based on cost.
FR222 = Enable customers to search based on type of sports.
FR223 = Enable customers to search based on other specifications like size or shape.

DP221 = Algorithm having provision to conduct cost-based search

DP222 = Algorithm having provision to conduct search based on sport
DP223 = Algorithm having provision to conduct search based on other specifications

The design equation at this stage can be written as follows:

$$\begin{bmatrix} FR221 \\ FR222 \\ FR223 \end{bmatrix} = \begin{bmatrix} X & 0 & 0 \\ 0 & X & 0 \\ 0 & 0 & X \end{bmatrix} \begin{bmatrix} DP221 \\ DP222 \\ DP223 \end{bmatrix}$$

FR3-DP3 decomposition

FR31 = Provide an excellent product check-out system.
FR32 = Transfer money from the purchaser.
FR33 = Protect security of customer's information.

DP31 = Shopping cart design
DP32 = Secure transmission system of money transfer
DP33 = Information encryption system

The design matrix corresponding to this decomposition is

$$\begin{bmatrix} FR31 \\ FR32 \\ FR33 \end{bmatrix} = \begin{bmatrix} X & 0 & 0 \\ 0 & X & 0 \\ 0 & X & X \end{bmatrix} \begin{bmatrix} DP31 \\ DP32 \\ DP33 \end{bmatrix}$$

FR4-DP4 decomposition

FR41 = Deliver goods within 5 business days or within 24 hours for urgent orders.
FR42 = Maintain capability to deliver within 24 hours.
FR43 = Deliver products in good conditions.

DP41 = Delivery system

DP42 = Inventory control of the products
DP43 = Good packaging system

$$\begin{bmatrix} FR41 \\ FR42 \\ FR43 \end{bmatrix} = \begin{bmatrix} X & 0 & 0 \\ X & X & 0 \\ 0 & 0 & X \end{bmatrix} \begin{bmatrix} DP41 \\ DP42 \\ DP43 \end{bmatrix}$$

FR41-DP41 can be further decomposed with transaction between company and postal services such as FedEx and UPS.

Similarly, we can perform other decompositions.

The following two examples are obtained from Nam Suh (2005) and Taesik Lee and Rajesh Jugulum (2003). We are thankful to Professor Nam Suh for his kind permission to publish these two examples.

7.5 DESIGNING A FAN BELT/PULLEY SYSTEM

Automobile engines have fan belts and pulleys to drive accessory equipment such as air-conditioning pumps, water-cooling pumps, and alternators. If the belts and the pulleys are not properly designed, the belt slips and makes undesirable noise. Let us design a fan belt/pulleys system that will satisfy the functional requirement of driving the accessory equipment without making any noise. Let us assume that the diameter of the pulley attached to the crankshaft of the engine is 7 inches and all other pulleys are 5 inches in diameter.

For the fan belt and pulley system (see Figure 7.5), the FRs and DPs may be written as follows:

FR = Drive accessory equipment without making noise.
DP = Fan belt pulley system

FR1 = Drive air-conditioning pump.

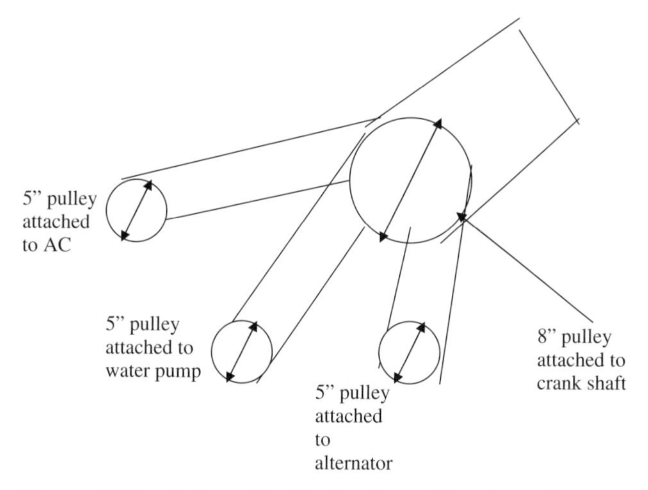

Figure 7.5 Axiomatically Designed Design of a Fan Belt/Pulley System.

FR2 = Drive water pump.
FR3 = Drive alternators.
DP1 = Belt and pulley connected to air-conditioning pump
DP2 = Belt and pulley connected to water pump
DP2 = Belt and pulley connected to alternators

In this design, we can have a noise elimination constraint.

The design equation may be written in the form of an uncoupled design:

$$\begin{bmatrix} FR1 \\ FR2 \\ FR3 \end{bmatrix} = \begin{bmatrix} X & 0 & 0 \\ 0 & X & 0 \\ 0 & 0 & X \end{bmatrix} \begin{bmatrix} DP1 \\ DP2 \\ DP3 \end{bmatrix}$$

We can further decompose these FRs and DPs. For example, let us decompose FR1 and DP1.

FR11 = Prevent slipping of belt sideways.
FR12 = Prevent slipping of belt along direction of motion.

The corresponding DPs are:

DP11 = Pulley design
DP12 = Frictional force between pulley and belt (The frictional force should overcome tensional force to prevent slipping in the direction of motion.)

The design equation at this level can be written as follows:

$$\begin{bmatrix} FR11 \\ FR12 \end{bmatrix} = \begin{bmatrix} X & 0 \\ X & X \end{bmatrix} \begin{bmatrix} DP11 \\ DP12 \end{bmatrix}$$

We can perform similar decomposition of other FRs and DPs.

7.6 USE OF DESIGN PRINCIPLES IN AN ACADEMIC DEPARTMENT*

In this section, it is shown how axiomatic design principles are used in an academic department to improve its overall functionality. This exercise was carried out by Professor. Nam Suh when he was the chairman of the Department of Mechanical Engineering at MIT.

7.6.1 Mechanical Engineering Department at MIT

The Department of Mechanical Engineering is one of the original departments that was established when the Massachusetts Institute of Technology (MIT) was incorporated in 1861 in Boston, Massachusetts. The department has approximately 60 faculty members, 750 students, and 50 technical and support staff members.

The department has been rated as the number one mechanical engineering department in the country ever since the game

*This example is from Suh (2005).

of rating academic departments started. It has produced some of the pioneering textbooks and monographs in a number of fields, including solid and fluid mechanics, thermodynamics, design, manufacturing, tribology, control, internal combustion engines, engineering analysis, system modeling, robotics, gas turbines, and others. It also has innovated, perhaps more than any other academic department of its kind, many new technologies being used in industry.

The Department of Mechanical Engineering has gone through a few transformations during the last fifty years. The modern mechanical engineering department that is deeply rooted in research, especially sponsored research, may be traced to the transitional period right after World War II. After the war, many outstanding faculty members came to the department from various institutions and departments.

The department experienced a turbulent period in the late 1950s and the early 1960s, due to several factors: the launching of *Sputnik* by the Soviet Union, the decreasing enrollment in the department, and demanding authorities to concentrate research in one particular area, ignoring fields such as design and manufacturing (that did not have a strong science base). Suh (2003) thinks that the department should have attempted to create the science base for these fields that did not have a science base rather than deemphasize them.

Beginning in the 1970s, it was realized that MIT must do more for the fields of design and manufacturing because the United States had its first trade deficit and U.S. companies were no longer dominating the manufacturing industry worldwide. In 1975, the department decided to create a major activity in the field of manufacturing. The department created the Laboratory for Manufacturing and Productivity (LMP) in 1976, which became an interdepartmental laboratory of MIT two years later. LMP concentrated on the two ends of the research spectrum: creation of the science base in design and manufacturing, and innovation of new technologies in polymer and metal processing. During this period, the department established strong links to industry by creating the first industrial consortium in the field of polymer processing and later in internal combustion engines.

In the early 1990s, there was a general consensus that the department has to reinitialize its research and educational activities. There was a feeling that it has been on a path of the time-dependent combinatorial complexity for too long. Beginning in 1991, a major effort was undertaken to redefine the discipline of mechanical engineering and make a significant impact on the knowledge base and technology innovation. At the beginning, it was not easy to embark on a new path, although Professor Suh had taken the department head job because the search committee convinced him that the department was ready for a change.

7.6.2 FR-DP Identification

Although this example was intended to explain complexity in academic units, we thought it would serve as a good case of explaining axiomatic design theory. When reinitialization was done, axiomatic design principles were used to satisfy all FRs. Please note that reinitialization is done to establish functions at the start of a new period and establish a best way to satisfy the FRs. This discussion should be part of complexity theory, where reinitialization is done to reduce one type of complexity. The details about this theory can be found in Suh (2003).

Changes were made to achieve the following three highest FRs of the department by reinitializing the department:

FR1 = Transform the discipline of mechanical engineering from one that is based on physics into one that is based on physics, information, and biology.

FR2 = Make impact through research on the knowledge base and technology innovation – the two ends of the research spectrum – rather than being in the middle of the research spectrum.
FR3 = Provide the best teaching to students.

The DPs that could satisfy these three FRs were chosen as follows:

DP1 = New research groups/efforts in information science and technology and in new bioengineering
DP2 = Shift in research emphasis
DP3 = New undergraduate curriculum

The PVs that could satisfy the DPs were as follows:

PV1 = Faculty members who can bring new disciplinary background into mechanical engineering
PV2 = Reward structure based on impact (rather than the number of papers or the amount of research funding)
PV3 = Gathering of financial resource to support new curriculum

7.6.3 Actions Taken

To achieve these goals (FRs), the following were done (not necessarily in chronological order):

1. Created the Pappalardo Laboratories to house new undergraduate laboratories in design, instrumentation, and project lab by converting and renovating 20,000 square feet of dilapidated laboratory space on the ground floor into a modern laboratory.
2. Created a new research laboratory in information science and technology – the d'Arbeloff Laboratory for

Information Systems – by renovating old space to house new research groups in mechatronics, automation of health care, and automatic identification of products.

3. The emphasis of bioengineering was changed from prosthesis to bioinstrumentation to create the "third leg" in the tripartite arrangement of medicine, biology, and engineering.
4. A new energy-related laboratory called the Laboratory for 21st Century Energy was created to develop the intellectual basis and technologies for the era when the demand for petroleum is greater than supply.
5. Hatsopoulos Microfluid Dynamics Laboratory was created by converting the traditional fluid mechanics laboratory.
6. AMP Teaching Laboratory in Mechanical Behavior of Materials was created by renovating the old laboratory.
7. A completely new undergraduate curriculum was adopted that offers integrated undergraduate subjects rather than traditional subjects to provide a better context for learning.
8. The Ralph E. and Eloise F. Cross CAD/CAM Laboratory was created for undergraduate teaching in design and manufacturing.
9. Cross student lounges were created for undergraduate students.
10. Since 1991, nearly 50 percent of the faculty members were replaced with new faculty members. Nearly 50 percent of these new faculty members came into the department from other disciplines such as physics, mathematics, optics, computer science, physiology,

materials, electrical engineering, and chemistry. Many of these new faculty members are now tenured, which should give continuity and permanence to the transformation of the department.

11. New chairs were created to recognize those who made special contributions.
12. Modern lectures halls, the B.J and Chunghi Park Lecture Halls, were created to accommodate new teaching methods to support the new curriculum and provide a better environment for learning and teaching.
13. A faculty prize for teaching innovation, called the Keenan Award for Teaching Innovation, was created that carries a reasonable stipend.
14. The Papplardo endowment fund for book writing was created to support the new textbook writing activities. Oxford University Press agreed to establish the MIT/Pappalardo Series for Mechanical Engineering and publish all the books written with the Pappalardo fund. A number of books have been published, and more will be published.
15. Many new research projects were created that are outside the traditional fields of mechanical engineering, such as two-photon microscopy for detection of cancer cells without incision, quantum mechanical computers, the use of the Internet and RF sensors to identify all the products, and others.

The department was successfully redesigned and reinitialized during this period. It is clearly the best department of its kind in the world.

7.7 DESIGNING A UNIVERSITY SYSTEM THAT WILL TEACH STUDENTS ONLY THROUGH THE INTERNET

This is one solution to the problem. There can be many solutions, depending on FRs identified.

FR = Teach university students using the Internet.
DP = University online teaching system

FR1 = Provide internet education to students.
FR2 = Teach students.
FR3 = Provide good technical and management support.

DP1 = Internet education course
DP2 = Internet support
DP3 = Management and technical support system

At this level, the design equation can be written in the form of a decoupled design:

$$\begin{bmatrix} FR1 \\ FR2 \\ FR3 \end{bmatrix} = \begin{bmatrix} X & 0 & 0 \\ X & X & 0 \\ 0 & X & X \end{bmatrix} \begin{bmatrix} DP1 \\ DP2 \\ DP3 \end{bmatrix}$$

FR2-DP2 decomposition

FR21 = Provide a course description.
FR22 = Supply training material.
FR23 = Develop plan for conduct of the course.
FR24 = Provide help sessions for students.
FR25 = Evaluate students' performance.
FR26 = Test students' ability to apply what was learned in the course.

DP21 = Course outline
DP22 = Course handouts

DP23 = Schedule of course
DP24 = Help session system
DP25 = Exams/HWs
DP26 = Projects

The corresponding design equation can be written as follows:

$$\begin{bmatrix} FR21 \\ FR22 \\ FR23 \\ FR24 \\ FR25 \\ FR26 \end{bmatrix} = \begin{bmatrix} X & 0 & 0 & 0 & 0 & 0 \\ 0 & X & 0 & 0 & 0 & 0 \\ X & X & X & 0 & 0 & 0 \\ 0 & X & X & X & 0 & 0 \\ 0 & X & 0 & X & X & 0 \\ 0 & X & 0 & X & 0 & X \end{bmatrix} \begin{bmatrix} DP21 \\ DP22 \\ DP23 \\ DP24 \\ DP25 \\ DP26 \end{bmatrix}$$

FR23-DP23 decomposition

FR231 = Provide Web-based lectures.
FR232 = Post teaching materials, HWs on the Web.

DP231 = Videotaped lectures
DP232 = PDF/Web files

The design matrix corresponding to this decomposition is

$$\begin{bmatrix} FR231 \\ FR232 \end{bmatrix} = \begin{bmatrix} X & 0 \\ 0 & X \end{bmatrix} \begin{bmatrix} DP231 \\ DP232 \end{bmatrix}$$

FR24-DP24 decomposition

FR241 = Provide a means to communicate between students and instructors and among students.
FR242 = Provide a means to have group communications.
FR243 = Provide a means by which instructor can give feedback.

DP241 = E-mail facility

DP242 = Chatting facility
DP243 = Bulletin boards

The design equation at this stage can be represented as follows:

$$\begin{bmatrix} FR241 \\ FR242 \\ FR243 \end{bmatrix} = \begin{bmatrix} X & 0 & 0 \\ X & X & 0 \\ X & X & X \end{bmatrix} \begin{bmatrix} DP241 \\ DP242 \\ DP243 \end{bmatrix}$$

Similarly, we can perform other decompositions.

Chapter 8

Implementing Lean Design

In the previous chapter on Lean and Six Sigma, we discussed the key principles behind lean thinking. The major emphasis is on *value* as defined by the customer. Lean approach has traditionally been used in improving existing processes by removing waste. We now know that about 70 to 80 percent of value as well as waste are created upstream during the design. Therefore, by proactively applying lean thinking during design phase, we can create better designs that prevent waste during its life cycle.

8.1 KEY PRINCIPLES OF LEAN DESIGN

The key principles of lean design is to design products, services, and processes with the objective of *preventing* waste, rather than reducing the waste during manufacturing or service delivery. The application of lean thinking falls into two categories: (1) to apply lean principles to improve the design process itself; and (2) to proactively create products, services, and processes that are lean by design.

The focus of the first approach is to identify and eliminate waste from the design process. This enables the design process to be agile and reduces the time to market new offerings.

The approach starts with mapping out the current design process and capturing the expectations from all stakeholders of the design process. We can now analyze the existing design process for waste and identify opportunities. Our next step is to create a future state of the design process. We then improve the process by eliminating the waste and minimizing non–value-added activities. In addition, we optimize the processes especially by identifying the process steps that can be executed simultaneously. A cautionary note is that the product development processes have certain characteristic that are unique, and therefore allowances must be made while we eliminate waste, standardize value-added steps, and manage the process.

The second focus of lean thinking in the context of design is creating designs that maximize value for the customer and have minimum waste opportunities built into them. The principle that we follow to achieve lean design is to maximize value and minimize cost and harm in the design for all the stakeholders. Therefore, we must first understand the stakeholders and all their value levers. Second, we must understand the events, processes and steps that deplete value for the stakeholders.

This concept is essentially the same as *ideality* from *Theory of Inventive Problem Solving*, also known as TRIZ. Ideality is the ratio of sum of all benefits to the sum of all costs and harm. Our objective during design is to improve ideality by improving the benefits and decreasing the costs and harm. As this function approaches infinity, it is called *ideal final* result. As long as the system has not reached Ideal Final Result, there are opportunities for improvement through waste elimination and defect reduction.

The numerator of ideality (benefits) represents the forces of values desired by the stakeholders. Stakeholders include customers, providers, regulatory agencies, environment, and society at large. At a minimum, we must understand the voice of key stakeholders such as customers and providers to capture the value dimensions desired by them.

So what are some of the key value dimensions desired by customers? To capture these dimensions, we must scan the life cycle of a product and every interaction of customer with the product during its lifecycle. Therefore, we must take into account of value levers during product acquisition process, product delivery process, product usage, product maintenance, and product disposal process. Many such value levers include ease of acquisition, affordability of the product, performance and perception expectations regarding the use of the product, maintainability of the product, durability of the product and disposability of the product.

Similarly, we can identify value dimensions desired by the provider. Again, we must consider the end-to-end process from product or service ideation to product or service disposal at the end of life. Some of the value levers include ease to manufacture; ease to repair; high barrier against imitation; low cost of manufacture, repair, maintenance, and disposal; higher levels of differentiation; low cost of investments; low risk; ease to market; low cost of product liability; and ease to leverage and create new products and services. Our objective is to capture these value dimensions from various stakeholders and maximize them.

The denominator of the ideality equation that we want to minimize is the cost and harm for all the stakeholders. These are also known as the *undesired outcomes* that lead to waste, and stakeholders want to avoid them. Following is a sample list of items that don't add value for stakeholders and hence should be avoided or minimized:

- High cost or difficulty of manufacturing or service delivery
- Many test and inspection requirements
- Difficult to maintain and repair
- Difficult to store or operate
- Complex to operate or use
- Sensitive to variation and noise from environment
- High skill required to use
- Detrimental to the environment
- Low availability and high cost for supplied materials
- Complex to manufacture, assemble, or deliver leading to high cost
- Expensive upfront investments in equipments
- Mandatory to have specialized knowledge, skill, or material

8.2 STRATEGIES FOR MAXIMIZING VALUE AND MINIMIZING COSTS AND HARM

So what are some common strategies we can adopt that will maximize value levers and minimize the value inhibitors? We list some common value levers and lean strategies to maximize them in Table 8.1.

Similarly, what are some ways to minimize the value inhibitors for all stakeholders? We list some of the common levers and strategies to minimize them in Table 8.2.

8.3 MODULAR DESIGNS

Many of the principles and approaches promoted by lean design thinking are covered elsewhere as a principle. For example, modular designs or architecture is a principle that is promoted by

Table 8.1 Value-enhancing Levers and Strategies for Maximizing Them

Value-enhancing Lever	*Maximizing Strategy*
Ease of product or service acquisition	Lean out the process steps so that customers can educate themselves about the product or service and easily acquire the product or service (just-in-time).
Performance level of supplied functions of the product or service	These items include quality, reliability, and other performance levers. Deliver on these expectations at the correct level. Undershooting and overshooting generates waste.
Ease of installation	Simplify installation process, create modular designs and mistake-proof designs.
Ease of operation	Create mistake-proof design; simplify operation approach.
Features of the product or service	Prioritize on customer expectations and ensure that we deliver options and features customers desire.
Perception and image	Develop the right image and perceived value of the solution.
High barrier against imitation	Obtain patent protection, copyright, trademark.
Leverage the design for new products or service	Use platform design, reuse concepts, and modular designs.
Ease of maintenance	Use modular design, access to subsystems.
Ease of disposal	Use disposable materials, modular assembly/disassembly.
Environmentally friendly	Design for environment.

Table 8.2 Value-inhibiting Levers and Strategies for Minimizing Them

Value Inhibitor	*Minimizing Strategy*
Cost to customer	Minimize the total cost of ownership, including acquisition, usage, and disposal.
Cost to provider	Standardize components, use low cost, high volume use and robust materials, automation, reduce costs for testing and inspection.
Complexity of manufacturing or service provisioning	Design for manufacturability or serviceability.
Sensitive to variation and noise	Use robust designs.
High test and inspection requirements	Design for Lean Six Sigma to create predictable design; improve design margins.
Heavy investments	Minimize dedicated equipment use, optimize tooling, jigs and fixture usage, process design to utilize existing processes and equipments.
High operational costs	Reduce handling, minimize transportation and movements, reduce consumable usage, and reduce WIP and inventory.
High design costs	Eliminate waste from design process, promote design reuse, optimize make/buy decisions, use existing design elements, platform designs, purchase off-the-shelf elements and subsystems.
High defects, scrap and waste	Design for Lean Six Sigma, process design, optimize design after piloting and prototyping, automate manual operations.
Specialized material and skill	Use low cost, reliable materials, and reduce part counts.

lean design thinking but can also be linked to axiomatic design. The first axiom on design independence promotes modular architecture. From a value creation perspective, module architecture promotes simpler interfaces that are easier to customize for multiple configurations, and provides flexibility for installation, transportation, maintenance, repair, and disposal.

The fundamental principle of modular design is to organize a complex system into a set of distinct subsystems that can be integrated or assembled easily to create the higher-level system. The principle applies to an engineering system such as mechanical device, electric circuit or nonengineering system such as organizational structure or service process design.

Modular designs are effective when the interfaces create an uncoupled design or decoupled design in the axiomatic design parlance. This minimizes the need and cost attached to system optimization. Effective modular design reduces the total cost for the provider because it enables the provider to reuse the subsystem design for other purposes. This is the basic thinking behind creating platform designs.

Design reuse can reduce development time and cost for the provider. For example, many software applications use a *login module* to allow the user to login into the system. If we create a module and associated subroutine for the login module, it can be reused again and again.

Modular approach enables us to easily optimize product or service performance at low recurring cost. For example, a decoupled or uncoupled brake system in a car allows us to optimize the vehicle's stopping function without having to change the design of other subsystems within the automobile.

If we utilize modular design principles and create platform designs, it enables us to create variant products with lower fixed and variable costs, since the incremental development costs are reduced through reuse of components or subsystem. Modular designs also improve the ability to install, test, maintain, repair, and service components or subsystems.

8.4 VALUE ENGINEERING

We have discussed in the previous paragraphs that the intent of lean design is to improve the *value* of the product, service, or process to be created. This raises the question of how value is defined, measured, and optimized. We have also indicated that value is defined by the stakeholders. Therefore, it is important to know who the stakeholders are, their role and power structure, as well as their expectations and perceptions of value. A common and simple definition of value is *performance generated per unit of cost*.

Therefore, we can deliver higher value by improving performance and keeping the cost the same or reducing cost for a given set of performance. This is the objective of *value engineering*. Value engineering has become a scientific approach or method in which we improve the product value by identifying, clarifying, and prioritizing the functions of our solution (product or service), relating these elements to the cost of delivering these functions, and optimizing the delivery of these functions at the lowest cost possible.

Otherwise stated, value engineering seeks to optimize performance by balancing cost and performance (functionalities). High performance at excessive cost and low performance at low cost are both unacceptable. Definition of value is tricky, but the most common approach is relating value to functions and performance. Value engineering is executed as follows:

- Define and scope the product, service, or process for value engineering analysis.
- Gather customer expectations regarding the product or service. These expectations can be performance or

perception expectations. Performance expectations are measurable and objective expectations. Perception expectations are ambiguous, subjective and difficult to measure expectations.

- Establish the functional requirements of the product or service.
- Prioritize the functions as primary or secondary. The objective of the prioritization is to ensure that performances of primary functions are not compromised and to ensure value is maximized.
- Establish design parameters and create a baseline design.
- Search for alternative design solutions without impacting the quality of the design (ensuring that design is decoupled or uncoupled and impact of variation on design is minimized).
- Generate design solutions that maximize value.

8.5 THE 3P (PRODUCTION, PREPARATION, PROCESS) APPROACH

The 3P approach popularized by Toyota aims to create lean designs and prevent future waste by simultaneously conducting the product design and the associated production process. Simultaneous product and process development will ensure that the production process can be established as a lean manufacturing system with the characteristics such as pull system, just-in-time inventory, minimum WIP, one piece flow systems, and flow lines and work cells.

- **Pull systems** – The production fulfillment is completed at the pull signal from customer. By eliminating waste and improving process speed, we can ensure that product manufacturing times are minimized, and the pull system enables minimum inventory and WIP. The 3P process supports building of pull system through design for assembly strategies, standardizing on fewer parts, and reducing the number of subassemblies.

- **JIT inventory system** – A lean enterprise minimizes the finished goods inventory and WIP (work in process) so that drain on capital, floor space, cash flow, material handling equipment, and additional labor are all minimized. A pull system supports the just-in-time (JIT) inventory model. The 3P process enables optimizing on suppliers, establishing JIT supply system, and implementing design for manufacturability strategies.
- **Work cell** – Once we establish a JIT system and pull system, we can now reduce waste even further by minimizing unnecessary movements and process motion. By creating tightly integrated flow-lines and work cell systems, we can improve the maturity of the lean enterprise. Again, the 3P process can be used to ensure design enables the use of existing process flow-lines and work cells.
- **One-piece flow** – The cycle time is improved through one-piece flow in favor of batch processes. This system enables minimizing of inventory and faster responsiveness to customers. With the help of 3P process, we can ensure that we select processes that support one-piece flow over batch processes.

Lean enterprise is achieved during production by the simultaneous progression of production process and product design phases. During each phase, information is exchanged between the product design and process design activities. As design concepts are generated, simultaneously processes needed to produce these concepts are evaluated. During the concept development and feasibility phase of the design, the production process design activities include product documentation, identification of major parts and suppliers, and rough estimation of the production process. During the preliminary design phase, multiple process design alternatives are considered. During the final design phase, production process is finalized with such details as takt time, cycle time, process capability, standard work, and production process layout.

Although lean design approaches evolved separately from Six Sigma approaches, these approaches have converged. For example, many practices such as modular designs, design reuse, and parallel development of product and process design were developed initially through lean thinking. They are validated now through axiomatic design and other frameworks. Lean Design and Design for Six Sigma both promote maximizing the function, which is the ratio of all benefits to all costs and harm. The convergence is the objective of Design for Lean Six Sigma.

Chapter 9

Theory of Inventive Problem Solving (TRIZ)

TRIZ is a problem-solving method based on creativity, logic, and data, which enhances the ability to solve these problems. TRIZ is the (Russian) acronym for the *Theory of Inventive Problem Solving*, which was developed by G. S. Altshuller and his colleagues in the former U.S.S.R. starting in 1946. Using TRIZ, one can identify creative solutions to the problem by the study of the patterns of problems and solutions. These patterns are discovered through the analysis of over three million inventions (or patents) to enable us to predict breakthrough solutions to problems.

9.1 INTRODUCTION TO TRIZ

Usually, TRIZ is used to accomplish the following:

- Create potential design solutions.
- Resolve design contradictions.
- Increase design options.
- Project technological path with various principles.
- Overcome psychological inertia.

TRIZ is being recognized and put into use in conjunction with Lean Six Sigma or Design for Lean Six Sigma (DFLSS) practices in organizational creativity and innovation initiatives. Development and subsequent research of TRIZ through the analysis of over 3 million inventions search has led to the following key discoveries:

- Problems and solutions were repeated across variety of applications with some specific patterns.
- Innovations that were analyzed used scientific effects outside the field where they were developed.
- These innovations led to about forty principles to overcome system conflicts (see Figure 9.1).

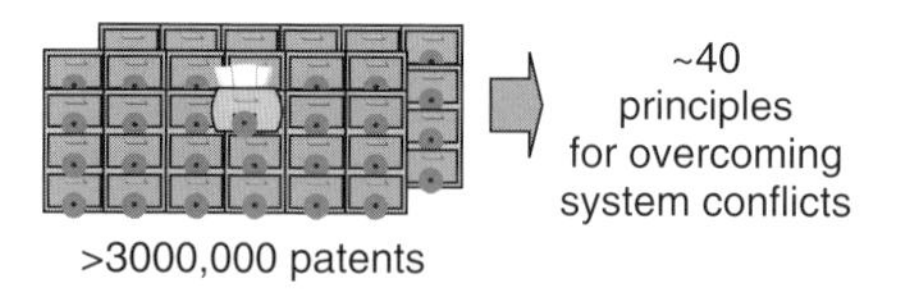

Figure 9.1 TRIZ Analysis of Major Innovations Leading to Key Inventive Principles.

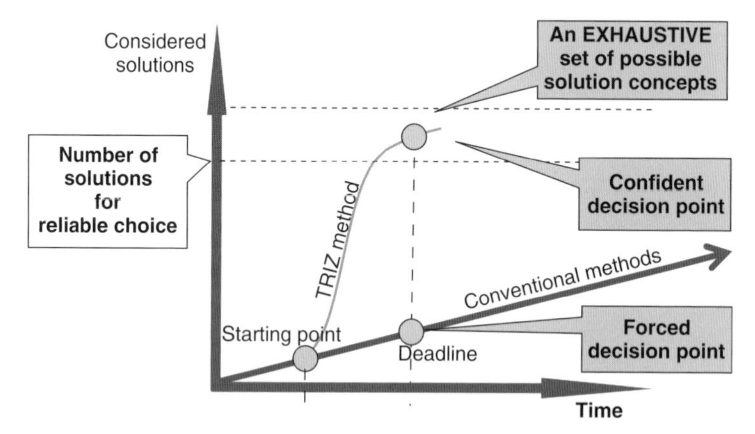

Figure 9.2 Impact of TRIZ on an Organization.

Impact of TRIZ in an organization can be explained clearly with Figure 9.2. In this figure, we can see how quickly we can reach a decision point when we have many solutions in a short period of time using TRIZ.

9.2 TRIZ JOURNEY

In contrast to general thinking, TRIZ makes the research effort accessible to the problem at hand. The problem at hand will be translated to a TRIZ generic problem. Using TRIZ tools, one can obtain a general solution that is finally translated into specific solution for the problem that is being addressed. These steps are shown in Figure 9.3.

9.2.1 TRIZ Road Map

Figure 9.4 shows a road map using TRIZ. After defining the problem, it is important to identify the ideal final result (IFR). Depending on the IFR, one can use TRIZ analysis tools and TRIZ database tools (if required). The new thing about this road map is the use of robustness principles in TRIZ database tools, to achieve robustness at concept level. Robustness can be briefly termed as insensitivity to noise factors. Robustness is achieved by making the design insensitive to variation due

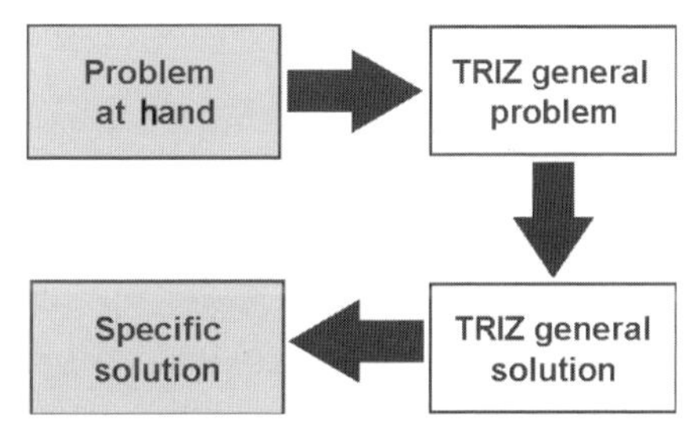

Figure 9.3 TRIZ Philosophy of Problem Solving.

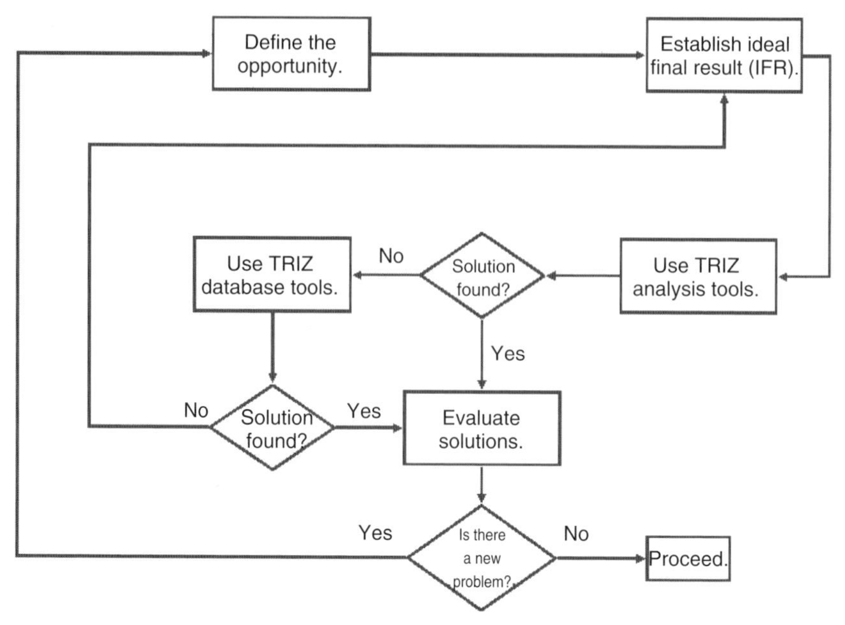

Figure 9.4 TRIZ Road Map.

to changes in process parameters, changes in usage conditions by the customer, and changes in product or material properties over time, or from repeated usage. Conceptual robustness is very important, since robustness can be considered to capture customer requirements at various usage conditions. In this chapter we provide a detailed discussion on these robustness principles.

As shown in the road map, the most important steps after problem identification is developing the ideal final result. An ideal system requires no material to be built, consumes no energy, occupies no space, and does not require any maintenance. It does not exist as a physical entity, but its function is fully performed. The characteristics of ideal system include the following:

- Occupies no space
- Has no weight

- Most robust
- No pollution
- Requires no labor
- Takes no time
- Requires no maintenance

Ideal final result (IFR) describes solution to a problem, independent of mechanism and constraints of the original problem. IFR can be developed from the following:

- Ideality equation
- Itself method

9.2.2 Ideality Equation

Ideality equation represents the degree of ideality, which can be defined as the ratio of functionality of the product (useful effects) and the cost plus harmful effects (side effects). Since no system is ideal in the real world, we can only measure the degree of ideality. A higher degree of ideality corresponds to the better design or solution. The ideality equation can be represented mathematically as

$$\text{Degree of ideality} = \text{Functionality (benefits)}/(\text{Cost} + \text{Harmful or side effects})$$

The evolution of ideal system is in the direction of increasing ideality (see Figure 9.5).

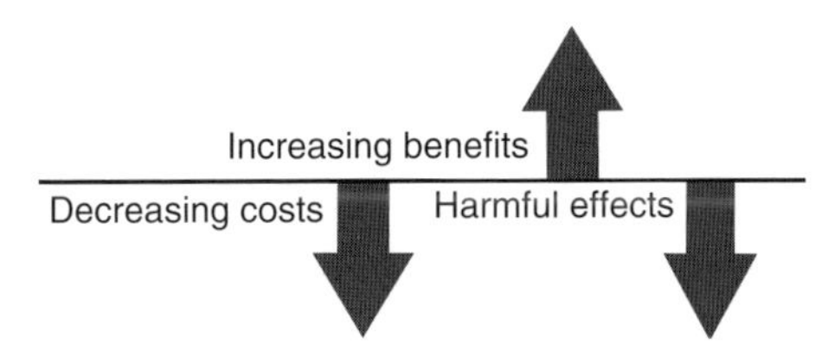

Figure 9.5 Increasing Ideality Leads to Innovation.

9.2.3 Itself Method

Another approach to develop an IFR is to express the ideal final result using the *itself* approach. The thought process here is that "the problem takes care of itself." Following are some examples for the itself method:

1. The grass mows itself – The grass keeps itself at an attractive height.
2. The data enters itself – The data stay accurate, current, and in correct form.

In the next two sections, we will briefly discuss TRIZ analysis tools and TRIZ database tools.

9.2.4 TRIZ Analysis Tools

Once the IFR is defined and the associated ideality is measured, then we proceed to conduct analysis of the problem at hand using TRIZ analysis tools. The most common TRIZ analysis tools are

- Resources
- Functional analysis
- Nine windows

Resources

The objective is to identify all the resources available to us inside the system as well as outside the system in order to achieve a higher state of ideality. Often, many of the resources are hidden from our view, not obvious or taken for granted. Many problems are solved through the recognition and straightforward use of these resources. Examples of some of these resources include air, vacuum, waste, and harmful substances and energy.

Typically resources are cataloged into the following categories:

- **Substance resources** – Any material from which the system or its environment is formed.
- **Field or energy resources** – Mechanical, electrical, chemical, thermal, magnetic, and electromagnetic field available.
- **Human resources** – Effective utilization of available people resources to conduct useful functions.
- **Time resources** – Resources for effective utilization of time. For example, we might be able to conduct parallel operations simultaneously. Another option is to utilize unused portions of time before, after, or during a process.
- **Information resources** – Information or data available from the system or people can be used for performing useful function.
- **Functional resources** – Potential of a system or its environment to perform certain function or tasks. These might include conduction, flotation, insulation, and reaction.

Many inventions were generated when designers realized that they could utilize unused energy, air, vacuum, and naturally available resources or waste from the system. For example, one of the electric power plants was trying to solve the problem of excessive secretion of selenium into the environment through the wastewater. Too much selenium is harmful for humans and can cause breathing difficulties or irregular heartbeat. The typical solution that involves the design of a system to separate the selenium and safely dispose of it was found to be very expensive. TRIZ analysis of resources revealed that a naturally occurring resource near the plant – namely, cattails and ragweed – can absorb selenium and bind it in their tissues. Therefore, the solution was to create a swamp near the exit of the power plant with the selenium-absorbing plants. In addition, they found that cotton and tobacco farms need selenium as a growth element, so they harvest the dead plants and sell them to the farmers.

Functional Analysis and Trimming

Many techniques that are closely related to Functional Analysis have been popular in the field of engineering for a while. These include Value Engineering or Value Analysis, Functional Cost Analysis, Value Stream Mapping and Analysis, Failure Mode and Effects Analysis, and Fault Tree Analysis. The objective of most of these approaches is to improve the performance of the system through analysis of various system, subsystem or component functions. Although experienced TRIZ users utilize Su-Field analysis to understand the functioning and behavior of the system, less experienced users conduct much easier functional analysis to analyze the elements of the system and its interactions with each other.

The first step in functional analysis is to describe each of the elements in the system, list the functions carried out by each of the elements. These functions are then grouped into useful functions, harmful functions, necessary functions, and insufficient functions. Our objective, then, is to find a way to improve the useful functions and minimize or eliminate the harmful and unnecessary functions.

Let's take the example of a medical syringe (Table 9.1) used for injecting medicine into patient's hand. First, we define the elements of the system and its environment under investigation. In this example, the system consists of syringe (cylinder, piston, and needle), medicine, patient's tissue, and nurse's hands. We then go on to describe the functions performed by each of these elements. They are described in Table 9.1. The next steps of analysis depend on the scope and objective of the problem solving

Table 9.1 Example of Medical Syringe

No	*Subject Element*	*Function*	*Object Element*	*Value Characteristic*	*Value Characteristic*
1	Cylinder	Holds	Medicine	Useful	
2	Cylinder	Guides	piston	Useful	
3	Piston	Pushes	Medicine	Useful	
4	Needle	Directs flow of	Medicine	Useful	
5	Needle	Hurts	Patient's tissue	Harmful	
6	Nurse's hand	Holds	Patient's tissue	Useful	
7	Nurse's hand	Holds	Cylinder	Useful	
8	Nurse's hand	Pushes	Piston	Useful	
9	Medicine	Penetrate	Patient's tissue	Useful	Necessary

at hand. For example, a simple redesign might only focus on eliminating harmful functions.

A common tool used to improve the design in connection with functional analysis is called *Trimming*. The objective of trimming is to eliminate unnecessary and harmful functions or improve inadequate functions. Key principles for trimming are as follows:

- An element in the system that does not provide a useful function or counteract a harmful function may be eliminated from the system.
- An element that provides a useful function and a harmful function may be eliminated from the system if the useful function may be provided by a remaining element in the system.
- An element in the system that does not contribute to the primary useful function must be evaluated concerning addition of value to the system.

We ask the following questions while conducting trimming of a function.

- Is the system able to operate without the function under investigation? If so, can we eliminate the subject performing the function?
- Can the object element perform the function by itself without the help from subject element?
- Can we use another element in the system to perform the function under investigation? Can we utilize any resource in the system or its environment to help us?

In our example, the value-added function is medicine penetrates patient's tissue. This is the function that we want to preserve or improve. The harmful function is "needle hurts patient's tissue." This is the function that we would like to eliminate. This function is caused by the needle. However, the needle performs a useful function: "Directs the flow of medicine into patient's tissue." Therefore, the question is whether some other elements in the system can carry out this function. The answer to this question depends on the scope of the project.

Nine Windows

Typically, we view the world through one "window" while the *nine windows* technique forces us to evaluate the world through nine different windows. Nine ways to think about the problem are created in the combinations of past, present, and future state, along with system, subsystem, and supersystem levels. A 3×3 matrix as shown in Figure 9.6 is used for our analysis. The multiple windows enable us to solve the problem at one or more of the nine ways. First, we look at the historical view of the problem from past, present, and future and its context. Another approach is to look at the system, the system with its environment (supersystem), and the details of the system at

	Past	Present	Future
Super system			
System			
Sub system			

Figure 9.6 Nine Windows.

lower level (subsystem). In combination, we ask if we can solve the problem at present, in preventive or in corrective fashion. In addition, we ask if the problem can be solved at a system level, supersystem level, or subsystem level.

Let's say the system under consideration is an aircraft gas turbine engine. Therefore we will place the engine under "Present System." At subsystem level, we will look at compressor, combustor, and turbine. We will place these under "Present Subsystem." In the supersystem, the engine is part of the aircraft. Therefore, we will place aircraft under "Present Super system." Now let's look the engine from past and future perspective. Engine in the past was at an assembly shop with its various subcomponents and subsystems. Therefore, we can place the unassembled engine in the "Past System." The same engine in an overhaul shop might fit in the category of "Future System."

If we look at subsystem level, we examine the compressor, combustor, and turbine. In the past, they were created from components using various machining, casting, and other means from raw materials. These would be placed under "Past Subsystem." In the future, we might look at degraded subsystems. For example, a combustor may have its liner needing replacement. It may be corroded. So we would look at the degraded combustor under "Future Subsystem."

Similarly, the airplane supersystem in the past was at an assembly line where wings, fuselage, engine, and control systems were getting ready for assembly. This is the view from "Past Supersystem." This supersystem in the future will be in an airline hanger getting ready for repair or overhaul. This is the "Future Supersystem" view.

Therefore, an issue facing the engine system could be solved at one or more of the nine window level. We could solve the engine problem in a preventive way before it happens at the engine level, subsystem (compressor, combustor, and turbine) level, or at the aircraft level. In a corrective way, we could solve the problem at these levels after the fact as well.

Similar analysis can be made of a business system as well. For example, we may have received complaints from our customer about an invoice (see Figure 9.7). We can treat it as a problem at the invoice (system) level, line item (sub system) within the invoice or as all the deliveries to customer (super system) level. Similarly, we can solve the problem in the present, in a preventive fashion or in a corrective fashion. Experience shows

	Past (Preventive)	Present	Future (Corrective)
Super system	Review line items before shipping.	Invoice line item	Correct line item errors.
System	Establish customer expectations about invoice.	Customer complaint about invoice	Apologize and correct invoice.
Sub system	Establish all expectations about purchase.	Customer delivery and payment system	Evaluate and improve all delivery systems.

Figure 9.7 Nine Ways to Solve a Customer Invoice Problem.

that major portions of the problems are resolved by using the TRIZ analysis tools. If the problem is not fully resolved, we resort to searching inside TRIZ database tools.

9.2.5 TRIZ Database Tools

TRIZ database tools were created as empirical principles derived from analysis of millions of inventions, including patents. The major database tools are contradiction matrix and separation principles, effects database, solution tree principles, and compilation of technological and business trends. These are graphically depicted in Figure 9.8.

Contradictions

One of the early findings of Altshuller (1984) was that obstacles to progress and innovation are the existence of contradictions. In addition, he observed that there are two types of contradictions – technical contradiction and physical contradiction.

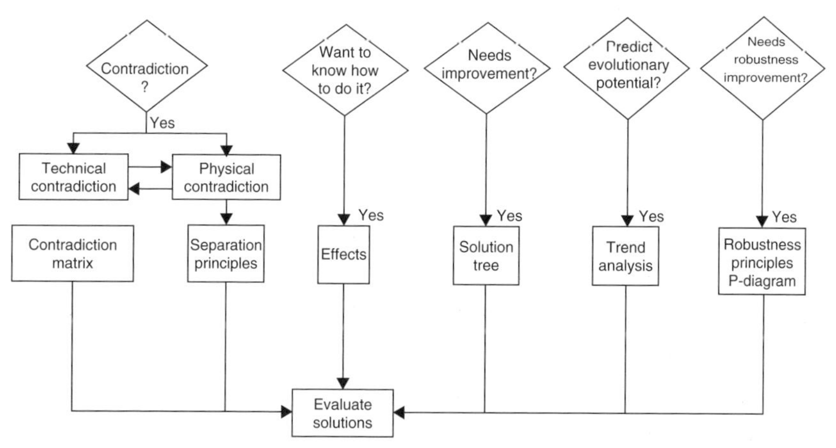

Figure 9.8 TRIZ Database Tools and Applications.

A technical contradiction is when we try to improve one parameter in the system, another parameter gets worse. For example, we would like to improve the weight of an object used for space flight application. Often, we can reduce the weight by using lighter materials. However, when we use lighter materials, it deteriorates the strength and integrity of the object. This would adversely affect the mission objectives of the flying system. Therefore, we have a technical contradiction between weight and strength of the materials.

Another example is that a business manager would like to review all the information relating to the department with the objective of getting better control of the departmental performance. However as the manager reviews more and more volumes of information, it adversely affect the time available for managing the departmental business. Therefore, as the parameter volume of information reviewed improves, it deteriorates another parameter – the time available for business management. Therefore, we have a technical contradiction between volume of information reviewed and time available for review.

The second type of contradiction is physical contradiction. Physical contradiction occurs when we want to maximize or increase a parameter for some reason and we want to minimize or reduce the same parameter for another reason. For example, we might want to increase the temperature of a mixture so that it can flow better through a small conduit. However, we don't want to increase the temperature of the mixture since it will decompose the mixture, which is undesirable. Here we have a physical contradiction of needing to simultaneously increase the temperature of the mixture as well as decreasing or maintaining the temperature of the mixture.

The resolution of technical contradiction is achieved with the help of a contradiction matrix devised by Altshuller. The matrix has a row of thirty-nine improving parameters, a column of thirty-nine worsening parameters, and a list of forty inventive principles. The thirty-nine parameters and forty principles are described in detail by several authors (example, Domb, E. and Tate, K. (1997), Domb, E. (1998), Silverstein et al., (2005)). They are also shown in Appendix A and Appendix B. At the intersection of each improving parameter and worsening parameter, we get a list of principles drawn from the forty inventive principles. They were derived from the analysis of inventive patents. The process for utilizing the matrix and the generating the solution is described in Figure 9.9.

Our first objective is to identify the contradiction in the problem at hand. Once the contradiction is identified, we must express it in terms of one the improving parameters and worsening parameters. In our previous example, we want to reduce the weight of the material used in flight. But this worsens the integrity of the system. So our technical contradiction is between weight and strength. Our next step is identify one of the thirty-nine improving parameters that resembles weight and one of the thirty-nine worsening parameters that resembles strength. Fortunately, one of the improving parameter is

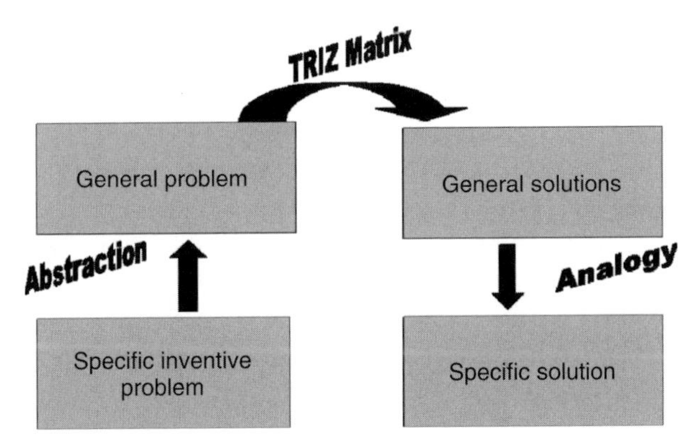

Figure 9.9 Generating a Specific Solution with the Help of TRIZ Contradiction Matrix.

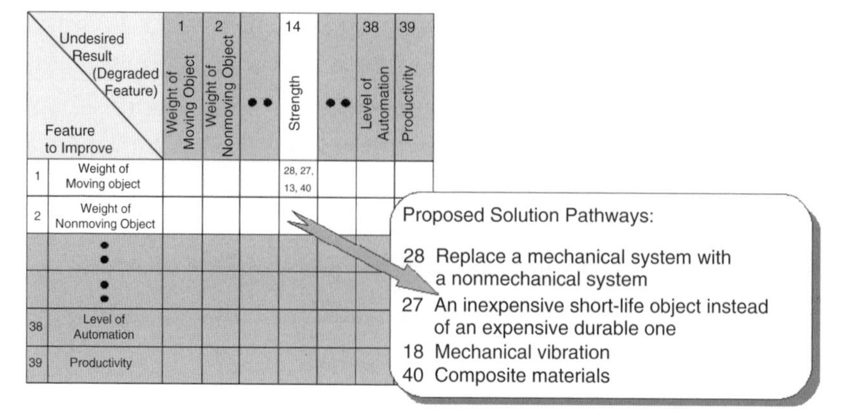

Figure 9.10 Use of Contradiction Matrix for Improving the Weight of Moving Object against Worsening Feature Strength.

"Weight of Moving Object" and one of the worsening parameter is "Strength." The intersection of the row "Weight of Moving Object" and the column "Strength" leads us to the principles 28, 27, 18, and 40 (see Figure 9.10).

Principle 28: Mechanics substitution

- Replace a mechanical means with a sensory (optical, acoustic, taste, or smell) means.
- Use electric, magnetic, and electromagnetic fields to interact with the object.
- Change from static to movable fields, from unstructured fields to those having structure.
- Use fields in conjunction with field-activated (e.g., ferromagnetic) particles.

Principle 27: Cheap, short-living objects

- Replace an inexpensive object with a multiple of inexpensive objects, with certain qualities (such as service life, for instance).

Principle 18: Mechanical vibration

- Cause an object to oscillate or vibrate.
- Increase its frequency (even up to the ultrasonic).
- Use an object's resonant frequency.
- Use piezoelectric vibrators instead of mechanical ones.
- Use combined ultrasonic and electromagnetic field oscillations.

Principle 40: Composite materials

- Change from uniform to composite (multiple) materials.

Our next step is to evaluate solutions 28, 27, 18, and 40. We would find that some of the solutions are more suitable and relevant for our application. In our particular situation, a composite material is an obvious choice.

In the second (business) example, we identified the contradiction between volume of information collected (improving parameter) and time available for review (worsening parameter). Our next step is to identify an improving parameter and a worsening parameter that resemble our situation. In this example, we might choose parameter 26, "Quantity of Substance/ Matter" or parameter 8, "Volume of Stationary Object" as the improving feature. Similarly, we might choose parameter 39, "Productivity" or parameter 25, "Loss of Time" as the worsening parameter. If we are not satisfied with one particular combination, we could explore all possible combinations. In this example, the intersection of improving parameter 26 and worsening parameter 39 produces principles 13, 29, 3, and 27. The next step is to carefully evaluate each of these principles and then generate specific solutions for our situation using analogy. The full matrix and 40 principles are given in Appendix A.

The resolution of the physical contradictions is achieved by the use of four separation principles:

1. **Separation in time** – Separate the necessity by establishing the property in the system at different time sequences. For example a chief engineer may complete a thorough and detailed review of a design only during critical design review times.
2. **Separation in space** – Separate the necessity by establishing the property in the system at different physical locations. For example, high security is established at locations where project teams are working on classified projects.
3. **Separation in scale** – Separate the necessity by establishing the property in the system at scales (macro versus micro). For example, a bicycle chain is rigid at the micro level and flexible at the macro level.
4. **Separation upon condition** – Separate the necessity by establishing the property in the system in different conditions. For example, an executive may conduct a detailed review of the project financials only if the budget exceeds a certain amount.

Let's consider an example of physical contradiction. An airplane needs wheels and landing gear for smooth takeoff and landings. However, wheels and landing gear are not desirable, since they increase the drag during flight. We want wheels, but we don't want wheels. Let's apply separation principles. The first principle says use the wheel and landing gear at certain times and avoid the using the wheels and landing gear at other times. In fact, this is the solution most airplanes use. Wheels and landing gear are tucked away during flight to produce a streamlined body. The same solution can be derived from other principles as well.

Effects

Another finding by the TRIZ researchers is that innovations that were analyzed used scientific effects outside the field where they were developed. Often times, we have exhausted all the solutions and methods relating the field in which the problem resides. According to the principle here, we may find an elegant solution to be borrowed from another field of study. For example, the health care industry was faced with the problem of collecting information from certain patients or during certain situations. The patient may need immediate help but may not speak the language of attending health care specialists or may be unable to talk when emergency procedure must be given. However, the attending health care professional may need such information as allergy information and previous medical history. The solution for the dilemma came from another field, financial services. It is common in the credit card industry to encode a person's pertinent information in the card so that it can be retrieved during a financial transaction. By borrowing from this solution, the health care industry is now able to create cards or bracelets with embedded chips containing patient information.

This approach is most useful when we are faced with a situation where no contradiction may exist but we still do not know how to achieve a certain result. In such scenarios, we look to other fields such as biology, chemistry, physics, business, mathematics, psychology, electrical engineering, chemical engineering, mechanical engineering, rheology, tribology, genetic engineering, and geometry. The most efficient way of identifying effects is through the utilization of *effects database* available in many TRIZ software. An alternate approach is to utilize the search engines available through the Internet.

Solution Tree

This technique is useful when the investigator is searching for improvements to the solutions. It is based on the seventy-six standard solutions compiled by Altshuller and his team, as well as the patterns of evolution to be discussed in the next section. The seventy-six standard solutions were grouped into five categories as described by Domb et al. (1999):

1. Improving the system with no or little change
2. Improving the system by changing the system
3. System transitions
4. Detection and measurement
5. Strategies for simplification and improvement

Technological and Business Trends

Another discovery by TRIZ scientists is that many patterns of technical evolution were repeated across industries and sciences. Early contributions to this field of research came from Altshuller. Altshuller discovered that the technical systems do not evolve at random, but they follow a certain pattern. He called these patterns *laws of evolution*. He went on to identify eight laws of evolution:

1. Technology follows a biological evolution of pregnancy, birth, growth, maturity, and decline.
2. Increasing ideality moves toward ideal final result.
3. Increasing dynamism and controllability occurs.
4. Increasing complexity is followed by simplicity.
5. There is evolution of matching and mismatching of parts.
6. There is a transition from macro level to micro level using fields to achieve better performance and control.
7. Nonuniform development of subsystems results in contradictions.

8. There is decreasing human involvement and increasing automation.

Subsequently, many researchers and authors have suggested many trends or laws of evolution, [Mann, 2002], the discussion of which are beyond the scope of this book.

9.3 CASE EXAMPLES OF TRIZ

In this section we present two examples where the TRIZ method was used with the help TRIZ analysis tools and database tools.

9.3.1 Improving the Process of Fluorination

The objective of this study is to improve the performance (uniformity) of the fluorination process. To accomplish the objective, it is required to find a source that would take the fluorine gas to each of the plastic bottles without increasing the cost and without making the system more complicated. TRIZ methodology is used to accomplish these objectives.

Fluorination Process

Fluorination process is a gas-modified plastic technology that reduces permeability and improves chemical resistance through surface treatment of the polymer. Fluorine gas is a strong oxidant that reacts with the plastic surface to replace the weak hydrogen molecule in the polymer. In other words, fluorination enhances the usability of a plastic container so that it can carry different solvents to help maintain product shelf life without solvent penetration. Basically, the cost for fluorinating any plastic surface outweighs the cost and usage of multilayer (an expensive resin), metal, and glass. Pesticides, two-cycle oil, mineral spirits, and solvents are just a few of the many solvents that are stored in fluorinated containers.

Customers are provided with four different types of fluorinated plastic bottles, depending on the application:

- B-24 – 20 to 40 percent of total surface uniformly fluorinated
- B-46 – 40 to 60 percent of total surface uniformly fluorinated
- B-68 – 60 to 80 percent of total surface uniformly fluorinated
- B-810 – 80 to 100 percent of total surface uniformly fluorinated

It should be noted that production of each type of these bottles is the same.

Application of TRIZ: Primary Function

We know that, fluorination of plastic bottles is usually done by introduction of fluorine gas in process reactor.

Ideality = (Sum of all benefits)/(Sum of all costs and harm)

Benefits

- Reduced permeability
- High chemical resistance
- Uniformity of fluorination on the plastic surface

Costs and harm

- Process cost of fluorination
- Cost, resources, and inconvenience of complex system

Ideal Final Result

Plastic bottles are fluorinated in a uniform manner over the desired surface with minimum amount of fluorine gas and using a simple system without the help from any additional resources.

Component List

Fluorine gas, process reactor, injection ports, conveying trolley, and plastic bottles are the components in this process.

Conflicting Components

- Article(s): Plastic bottles
- Main tool(s): Fluorine gas

System Conflicts

- (SC-1) Increasing the amount of fluorine gas will enhance the uniformity of fluorination in plastic bottles, but it will increase the process cost substantially.
- (SC-2) Decreasing the amount of fluorine gas will not enhance the uniformity of fluorination among plastic bottles, but it will decrease the process cost. This makes the process more competitive.

Conflict Intensification

- (SC-1) Infinite amount of fluorine gas will fluorinate the plastic bottles with high uniformity, but it will increase the process cost infinitely.
- (SC-2) Zero amount of fluorine gas will not fluorinate the plastic bottles, but it will reduce the process cost to a minimum.

Mini Problem

It is required to find such an X [source] that would take the fluorine gas to each of the plastic bottle without increasing the cost and without making the system complicated.

Conflict Domain

The area occupied by plastic bottles and fluorine gas is the conflict domain.

Operation Time

- **Pre-conflict time** – Before putting the plastic bottles in process reactor
- **Conflict time** – When the fluorine gas will be introduced in the process reactor
- **Post-conflict time** – After completion of fluorination

Resources

- Fluorine gas
- Process reactor
- Injection ports
- Conveying trolley
- Plastic bottles
- Gravity
- Vacuum

Use of Existing Resources

We can utilize the assistance of gravity available in the system since fluorine gas is heavier than air.

Technical Contradiction

Improving function: Uniformity of fluorination
Worsening function: Cost of fluorination
Improving parameter for the TRIZ matrix: Parameter 29 (Manufacturing precision)
Worsening parameter for the TRIZ matrix: Parameter 23 (Loss of substance)
Solutions: Principles 35, 31, 10, 24

Principle #35: Parameter Changes

Ideas: Will change in fluorine's or plastic bottle's parameters affect the fluorination process? This can be temperature, state of matter (gas, liquid, solid), or concentration.

Principle #31: Porous materials

1. Make an object porous or add porous elements (inserts, coatings, etc.).
2. If an object is already porous, use the pores to introduce a useful substance or function.

Ideas: Fluorination is essentially the process of introducing a useful substance to make the plastic bottle more usable.

Principle #10: Preliminary action

1. Perform, before it is needed, the required change of an object (either fully or partially).
2. Prearrange objects such that they can come into action from the most convenient place and without losing time for their delivery.

Idea: Is there any pretreatment of plastic bottle surface that can enhance uniformity of fluorination?

Principle #24: Intermediary

1. Use an intermediary carrier article or intermediary process.
2. Merge one object temporarily with another (which can be easily removed).

Idea: Use an intermediary object to improve the uniformity? We can use a blower to improve the local uniformity.

Solution

The process reactor is under vacuum during the reaction (0–10 Torr pressure). Since fluorine gas is 1.5 times heavier than air, it always tends to move downward due to gravitational force. If we position the fluorine gas molecules on the top of the plastic bottles, it will automatically move down. This will not increase the cost and complicate the system. The gas dynamics is an important property of the gas distribution. Therefore, by placing showerhead injection ports at the top of the process reactor, one can generate better gas dynamics and hence uniform distribution. This will enhance the reaction to achieve an optimal solution to the distribution problem.

Therefore, the ports are placed on the top of the reactor. Earlier, the ports were on the sides of the reactor. For improving the uniformity, a blower was kept below the ports. With this arrangement, the uniformity was greatly increased because of the gas dynamics properties.

9.3.2 Coordinate Measuring Machine (CMM) Support Problem

The objective of this study is to develop an innovative concept for the universal support by using TRIZ. This is necessary to solve a very common problem in the industry – supporting various kinds of parts of plastic or metal during the process of inspection.

Application of TRIZ: Primary Function

Different types of supports are used to align the components on the bed of a CMM to accurately perform various types of measurement operations.

Ideality = (Sum of all benefits)/(Sum of all costs and harm)

Benefits

- Support the part on CMM bed for inspection
- Easy and accurate measurement
- Simple and flexible system capable of holding variety of parts

Costs and harm

- Cost of support system
- Multiple support systems
- Setup time
- Damage to the part for inspection

Ideal Final Result

Parts to be inspected on CMM are held parallel to the CMM bed accurately by a simple support system. This system is solid enough to hold the part but flexible enough to take the shape of the part.

Component List

CMM bed, supports, parts, measurement probes are components in this process.

Conflicting Components

- Article parts to be measured
- Main tool CMM bed, supports

System Conflicts

- SC-1: Different types of supports can support one type of component parallel to surface of machine bed.
- SC-2: Same supports cannot support a different component, which makes the system complicated and inaccurate.

Conflict Intensification

- ISC1: Infinite types of supports can support virtually any shape, but complicate the system infinitely.
- ISC2: Zero support simplifies the system but cannot support the parts at all.

Mini Problem

It is required to find such an X [source] that would

- Support the parts parallel to CMM bed accurately
- Not make the system complex

Conflict Domain

Area of contact between the part and supports are the conflict domain.

Operating Time

- **Pre–conflict time** – Before putting a part on machine bed
- **Conflict time** – When a part will be placed at the top of supports
- **Post–conflict time** – After completion of inspection

Resources

CMM bed, supports, parts, measurement probes, air, gravity, magnetic field, and atmospheric pressure are resources.

Selection of X Resource

Atmospheric pressure

Technical Contradiction

Improving function: Versatility of support system to hold any part parallel to CMM bed

Worsening function: Complexity of support system

Improving parameter for the TRIZ matrix: Parameter 35 (Adaptability or versatility)

Worsening parameter for the TRIZ matrix: Parameter 36 (Device complexity)

Solutions: Principles 15, 29, 37, 28

Principle #15: Dynamics

1. Allow (or design) the characteristics of an object, external environment, or process to change to be optimal or to find an optimal operating condition.
2. Divide an object into parts capable of movement relative to each other.
3. If an object (or process) is rigid or inflexible, make it movable or adaptive.

Principle #29: Pneumatics and hydraulics

1. Use gas and liquid parts of an object instead of solid parts (e.g. inflatable, filled with liquids, air cushion, hydrostatic, hydro-reactive).

Principle #37: Thermal expansion

1. Use thermal expansion (or contraction) of materials.
2. If thermal expansion is being used, use multiple materials with different coefficients of thermal expansion.

Principle #28: Mechanics substitution

1. Replace a mechanical means with a sensory (optical, acoustic, taste or smell) means.
2. Use electric, magnetic and electromagnetic fields to interact with the object.
3. Change from static to movable fields, from unstructured fields to those having structure.
4. Use fields in conjunction with field-activated (e.g. ferromagnetic) particles

Physical Contradiction

Maximizing function: Supporting grip
Minimizing function: Supporting grip

Elimination of the physical contradiction: Separation of opposite properties in time:

- Atmospheric Pressure must be solid in the vicinity of the part's profile.
- It must become flexible in the vicinity of part's profile during conflict time and within conflict domain.

Solution

Two possible solutions to this problem are proposed.

Solution 1

We should have a mechanism that will behave like a solid rock by using atmospheric pressure during the conflict time and within conflict domain. The mechanism should become flexible after the completion of operation. Such an arrangement is described next:

If we enclose sand (Silica) in an elastic bag and evacuate the air inside the bag without removing the sand, the atmospheric pressure will be applied to the grains of sand to make it a solid support. By introducing the air again, the sand will become flexible. We can make this type of arrangement at the conflict domain only, as we normally do not require to support the entire area of the part to be inspected. The details are clearly depicted in Figure 9.11.

Solution 2

Another solution is as follows: We can use small steel balls instead of sand. We can apply the pressure (mechanical or atmospheric) on balls to make them solid. To make the bonding

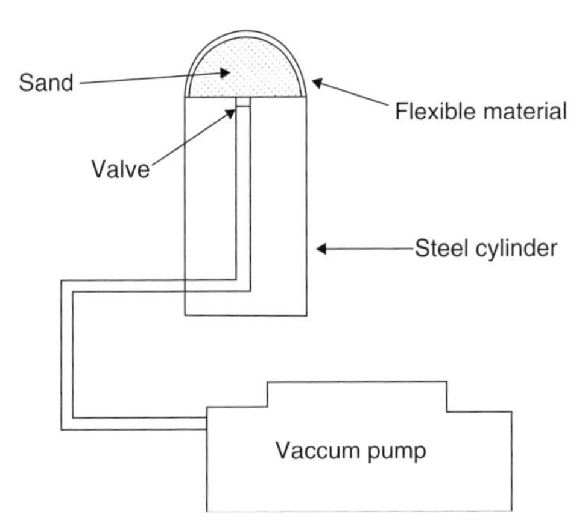

Figure 9.11 Solution 1 for Universal CMM Support.

of balls stronger, we can enhance the solidification stage by introducing the magnetic field at the conflict domain and during the conflict time.

Applications

This innovative support has number of applications in the industry. Three of the applications are as follows:

1. Supporting parts on CMM bed
2. Supporting parts on CNC machining center's bed
3. Supporting parts on a granite table for manual inspection

9.4 ROBUSTNESS THROUGH INVENTIONS

In this part of this chapter, we provide a discussion about concept design principles for improved robustness. These principles are known as P-diagram principles in TRIZ road map. They can be considered add-on TRIZ database of inventions as well as TRIZ principles. The goal of this research effort was to identify the principles that will improve robustness at concept level.

Robust design is a set of engineering methods for attaining high-quality function despite variations due to manufacturing, the environment, deterioration, and customer use patterns. The fundamental principle of robust design is to improve the quality of a product by minimizing the effects of variation without eliminating the causes [Phadke, 1989]. There are three generally recognized phases in the robust design process. Table 9.2 lists the three major phases of the design process emphasized by Taguchi (1988) – concept design, parameter design, and tolerance design. For each phase, some design activities are listed that have a major impact on robustness. These activities are listed in the second column of Table 9.2, roughly in chronological order. The strategies developed through this research can be effectively

Table 9.2 Phases in the Design Process and Design Activities Related to Robustness

Design Phase	*Design Activities Related to Robustness*
Concept design	Generate concepts to create the desired function
	Generate concepts to make a function more robust
	Evaluate concepts
	Select from among a set of concepts which ones to pursue
Parameter design	Plan a search through the design space
	Conduct experiments
	Analyze data
Tolerance design	Estimate the economic losses due to variations
	Allocate variations among components
	Optimize trade-offs between cost and quality

used in the first phase to generate concepts to make a function more robust.

Robust parameter design and tolerance design methods have made a significant impact on the quality of products and the speed with which they are developed. The tools for accomplishing robustness at the concept design stage appear to be less well developed. This project is intended to improve tools for the concept design stage. Specifically, we hope to provide tools to help engineers develop robustness inventions. The term *robustness invention* will be defined and explored in the next section.

9.4.1 What Is a Robustness Invention?

Robustness Invention

A *robustness invention* is a technical or design innovation whose primary purpose is to make performance more consistent despite the influence of noise factors. The patent summary and prior art sections in patent or invention description usually provide clues to find robustness innovation.

Every patent must describe its advantages over the prior art. The majority of patents are developed to provide new functionality, higher levels of performance, lower cost, longer life, or reduced side effects. But a substantial number of patents cite *robustness* as the principal advantage of the invention. This book will use the term *robustness invention* to refer to those patents whose primary advantage is that they are less sensitive to noise factors in the environment, customer use, or manufacture.

Patents on robustness inventions usually don't include the term *robust* – instead, they use terms such as "less sensitive," "more consistent function," "repeatable," or "regardless of changes in (name of noise factor)." The term *robustness* is not always used because many inventors aren't familiar with Taguchi methods, but they do have an appreciation for the value of consistent function under realistic conditions. Although robustness inventions are not always simple to identify directly by key words, they can be screened by key word and then identified manually by reading the technical descriptions in detail.

This section has defined robustness inventions and some of their hallmarks that allow them to be identified. The next section will describe our research methodology in which large numbers of robustness inventions are identified and analyzed.

9.4.2 Research Methodology

To accomplish our goal of accelerating robustness invention, the methodology that was used is similar to that employed by Altshuller [1984] in developing the Theory of Inventive Problem Solving, or TRIZ. Altshuller screened several patents looking for inventive problems and how they were solved. He ultimately came to the conclusion that patents can be best classified on the basis of how they overcame *contradictions* in previously existing engineering systems. It was found that Altshuller's research approach to be of great value. This research approach is composed of four steps:

1. *Collect a large body of inventions related to robustness.* Patents from searchable full text databases, such as the U.S. Patent and Trademark Office on-line database were drawn.
2. *Analyze the inventions in detail to understand how they achieved improved robustness.* Use of engineering models of the inventions whenever needed to identify the basic working principles involved.

3. *Seek useful classification principles for the inventions.* This is the most challenging aspect of the research. The principles are classified based on parameter diagram (P-diagram).
4. *Develop a database organized around the classification principles.*

9.4.3 Results of the Patent Search

In order for the research methodology described in the previous section to be viable, there must be a large number of patents whose principal advantage is a robustness improvement. The U.S. Patent and Trademark Office (USPTO) has issued nearly 7 million patents, which is a dauntingly large body of well-documented innovations. However, most of these are not robustness inventions. To carry out this work, there was a need to call out a reasonably large set of robustness inventions from this larger set. A key word search gives us some sense of the viability of this enterprise. The USPTO has a full-text, searchable database of all the patents issued back to 1976, over 3 million patents. Table 9.3 shows the results of some database queries looking for terms related to robustness in the abstract and body of the patents. Less than half of the patents retrieved have been confirmed to be robustness inventions according to the definition of this paper. Nevertheless, we estimate that there are more than 100,000 U.S. patents that meet our definition of robustness invention.

9.4.4 Robust Invention Classification Scheme

Based on a small sample of patents analyzed so far, we have developed a preliminary scheme for classifying robustness inventions. This scheme should be regarded as a rough draft, since we

Table 9.3 Key Word Search for Robustness Inventions

Search Term	*Number of Hits*
Insensitive	35,708
Less sensitive	12,253
Robust	27,913
Accurate	221,600
Reliable	211,533
Repeatable	16,458
Tolerant	13,765
Despite changes	1,323
Regardless of changes	1,147
Independent of	20,521
Self compensating	1,269
Force Cancellation	59
Independent	114,201
Uncoupling	2,189
Decoupling	6,505
Noise compensation	22,092
Noise control	142,138
Noise conditioning	10,787
Resistant	3,535
Acclimation	712
Desensitize	447
Sweet spot	1,317
Operating window	728
TOTAL	**867,472**

have studied less than 1 percent of the available patents. The scheme may change on the basis of what we learn as the project progresses.

The first layer of our taxonomy for classifying robustness inventions is based on the concept of a P-diagram or parameter

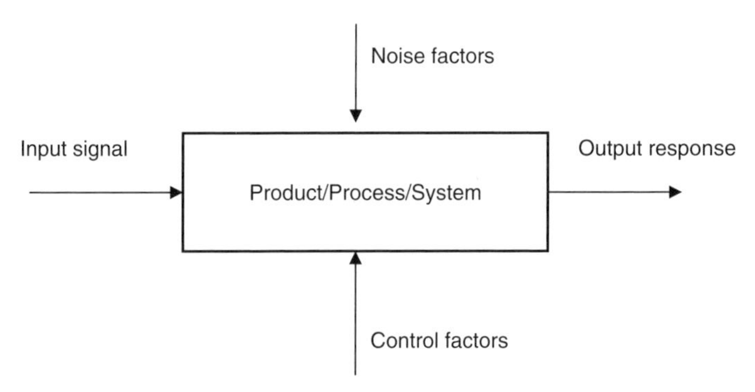

Figure 9.12 Parameter Diagram, or P-diagram.

diagram. The strategies are organized into four major classes based on a P-diagram, as depicted in Figure 9.12. A P-diagram is a block diagram used to represent a system depicting its input signal, control factors, noise factors, and response.

We may sort any robustness patent according to whether the gain in robustness was achieved by acting on the signal, the response, the noise, or the control factors. Examples of each of these major classes of robustness inventions are provided in Figure 9.13. Based on P-diagram elements, nineteen strategies were identified based on the analysis of about two hundred inventions.

9.4.5 Signal-based–Robust Invention

One class of robustness inventions works principally by acting on the signal. Two examples are provided in this section.

Selective Signal Amplification

In this class of patents, the system is reconfigured to amplify effects of the signal factor, while not appreciably amplifying the effects of the noise. Thus, one may make substantial improvements in the signal-to-noise ratio.

An example of this class of inventions is patent #5,024,105 (see Figure 9.14). In the rotameter of the prior art, flow rate is

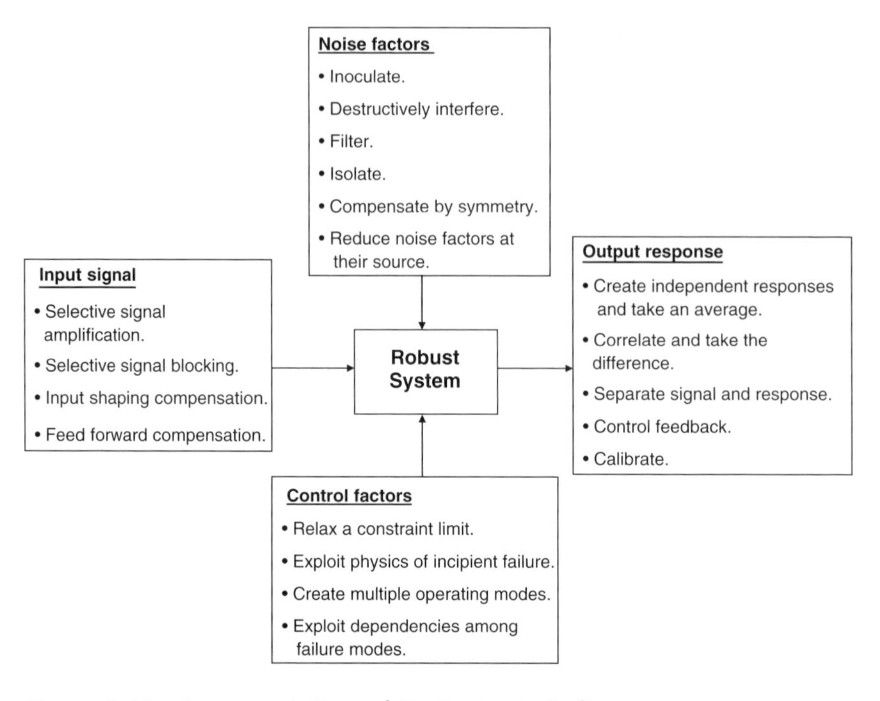

Figure 9.13 Representation of Strategies in P-diagram.

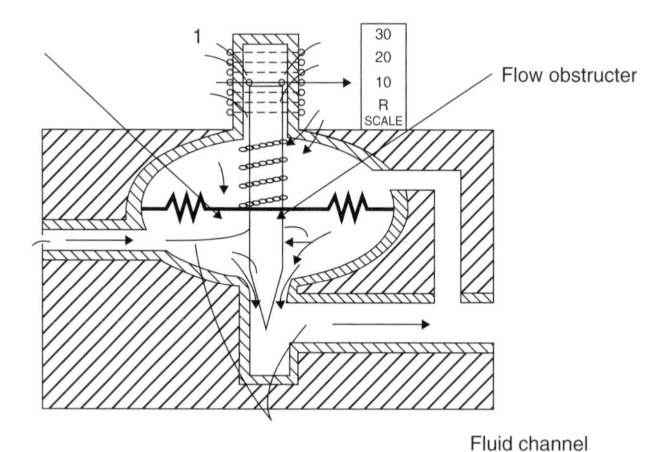

Figure 9.14 Viscosity-insensitive Variable-area Flowmeter.
Source: Adopted from Tentler and Wheeler (1991).

measured by forcing fluid to flow through a variable area duct, which causes a pressure change proportional to flow rate. But changes in the viscosity of the fluid (the noise) cause drag, which adversely affects the measurement. In this design, the pressure change is magnified by allowing it to act on a large surface. Viscous forces are not amplified by this design because the flow is mostly parallel to the surface. Thus, ratio of pressure force to viscous force (signal to noise ratio) is improved substantially.

Selective Signal Blocking

In this class of patents, features are added to the system that cause the signal to be ignored under certain conditions. This may be necessary when the consequences of the noise are severe within a certain identifiable range of conditions in which the system becomes hypersensitive to noise. Under these extreme conditions, it may be advantageous to eliminate the effects of both by blocking both signal and noise.

An example of this family of patents is #5,627,755, "Method and system for detecting and compensating for rough roads in an anti-lock brake system." The signal to the system is acceleration of the wheel and the normal response is to release brake pressure to prevent wheel skidding. The noise is wheel acceleration caused by rough roads. In this invention, a desensitizing factor is applied when rough road conditions are detected. In effect, both the signal and the noise are blocked under a prescribed set of conditions in order to render the response insensitive to rough road conditions (see Figure 9.15).

9.4.6 Response-based Robust Invention

This major class of robustness inventions works principally by acting on the response directly to reduce sensitivity without eliminating the cause. An example is provided below.

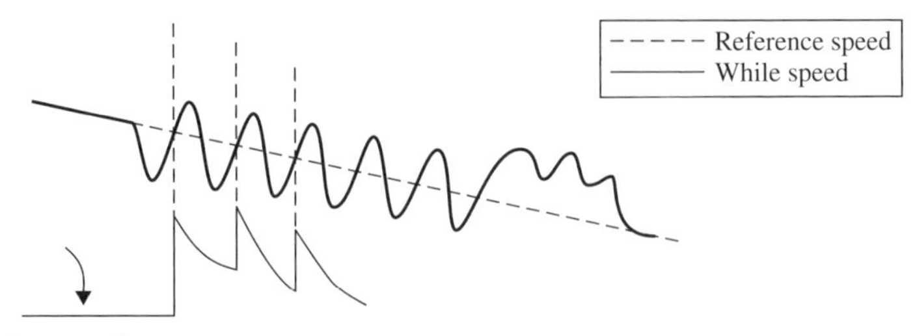

Figure 9.15 Method and System for Detecting and Compensating for Rough Roads in an Anti-lock Brake System.
Source: Adopted from D. Negrin (1997).

Correlate and Take a Difference

In this class of patents, the system is reconfigured to create two responses to the signal. The two responses must both be affected by the noise so that the two effects are strongly correlated, equal in magnitude, and opposite in sign. By taking a difference between the two responses (either electronically or mechanically), we create a response that is substantially insensitive to the noise.

An example of this category is Patent #5,483,840 "System for Measuring Flow." Here the velocity of a fluid at the core of a pipe is to be measured by means of the drag induced on a body on the flow (see Figure 9.16). However, the fluid is highly non-Newtonian, so its viscosity varies with shear rate, as well as temperature and fluid composition. The signal is the fluid velocity at the core of the pipe. The noise is the viscosity, which changes with shear rate. The invention resolves the problem by placing a disc in the flow with one edge near the core of the pipe and the other edge near the wall of the pipe. The disk will rotate at exactly the rate that makes the shear on both sides balanced. That condition will occur when the relative velocities of the disk and the flow are equal, regardless of the viscosity of the fluid. In

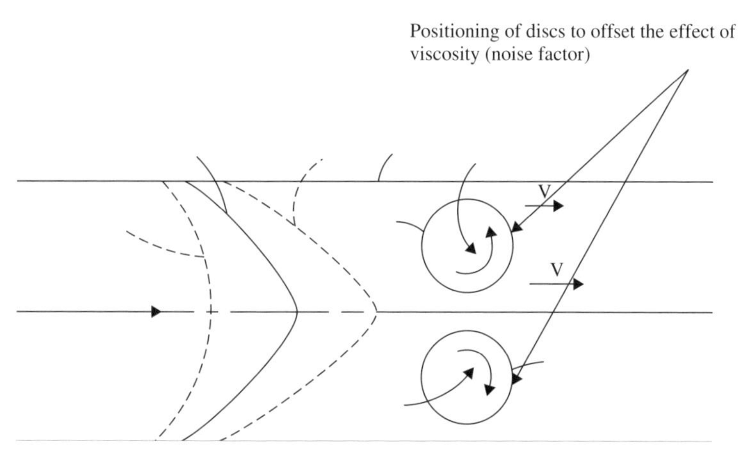

Figure 9.16 System for Measuring Flow.
Source: Adopted from Victor Chang, O. Chang, and M. Campo (1996).

effect, the disk takes a difference between two quantities (shear forces) and equates that difference to zero. Both sides of the disk are affected to the same degree by the noise factor. Hence, the rate of rotation of the wheel is a function of only the velocity of the fluid and is unaffected by viscosity of the fluid.

9.4.7 Noise-factor–based Robust Invention

One class of robustness inventions works principally by acting on the noise. Some obvious approaches include insulating the system from noise or filtering out noise. The less obvious approach of inoculating a system against noise is described in this section.

Inoculate against Noise

In this class of patents, a dose of the noise factor applied to the system renders the system less sensitive to the very same noise factor. One example of this category is Patent #4,432,606 "Optical fiber insensitive to temperature variations." Optical

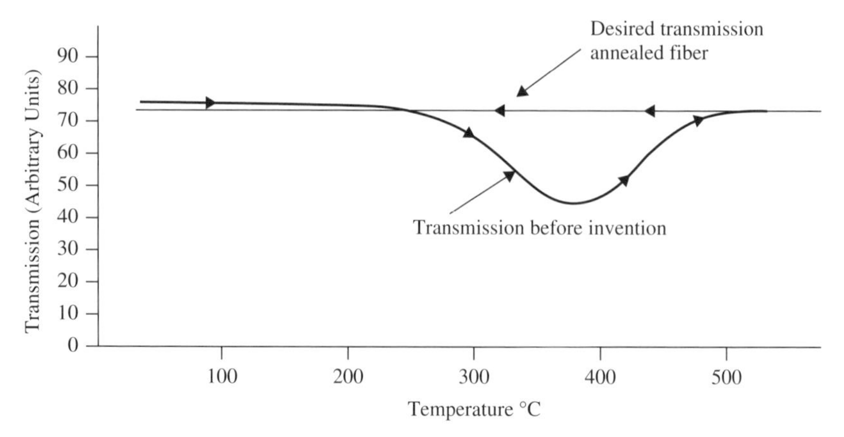

Figure 9.17 Optical Fiber Insensitive to Temperature Variation. *Source:* Adopted from R. Blair (1984).

fibers have been found to evidence a substantial decrease in optical transmission as a function of increased temperature (see Figure 9.17). In this patent, annealing a metal-coated optical fiber at 560 centigrade is found to render the fibers insensitive to temperature variations between about −200 at least about 560 degrees C.

Compensation by Symmetry

A second example in this class is related to the principle *compensation by symmetry*. The example is the material invar, for which Charles Edouard Guillaume was awarded the 1920 Nobel prize in physics. The dimensions of a piece of invar are substantially robust to temperature variations within a limited range of temperature (see Figure 9.18). This is brought about by setting the Curie temperature of the alloy by means of the relative proportions of nickel in the iron. At low temperatures the spins of the electrons all point in the same direction, making the alloy ferromagnetic. As the temperature is increased, however, the spins start to point in random directions and the volume of

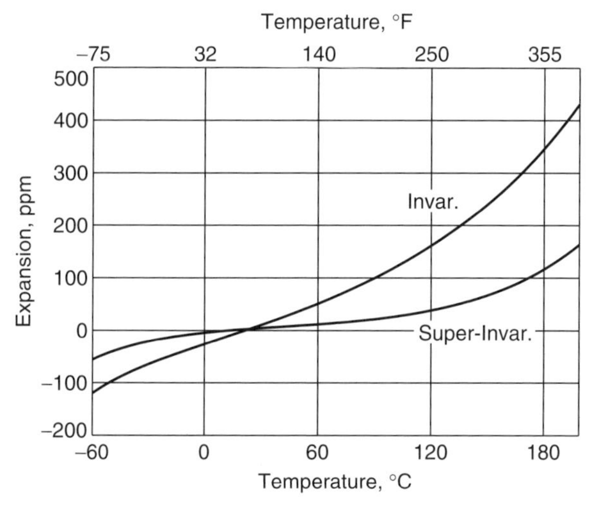

Figure 9.18 Invar Alloy.

the unit cells decrease. This reduction in volume compensates for the expansion caused by increased thermal vibrations. This creates a local plateau in the length versus temperature curve. Many measuring devices include invar to make their function more robust to temperature fluctuations.

9.4.8 Control-factor–based Robust Invention

The control factors in the P-diagram represent physical parameters whose nominal values can be selected by the designer. In robust parameter design methods, such as approaches pioneered by Taguchi, the designer systematically explores control factor settings while simultaneously inducing noise factor variations in an effort to make the system response relatively insensitive to noise factors. In concept design, there are strategies by which actions upon the control factors can either make the system inherently more robust or make later parameter design steps more effective. We have identified four strategies as shown in

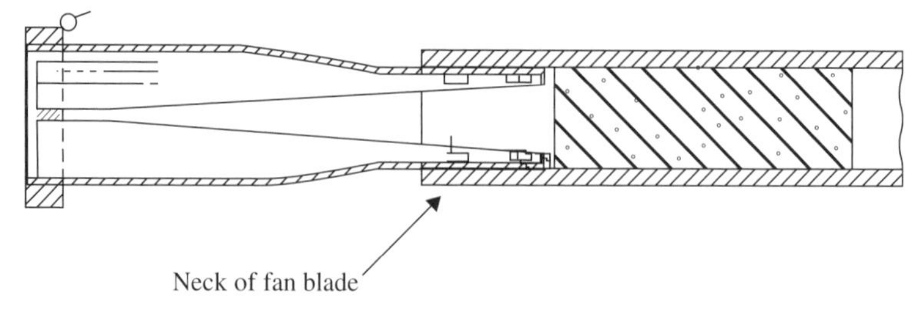

Figure 9.19 Axial Flow Fans and Blades Therefore.
Source: Adopted from R. C. Monroe (1982).

Figure 9.13 in which changes in control factors are the primary means of improving robustness.

An example of in this class of inventions is #4,345,877, "Axial flow fans and blades therefore." This invention is related to redesign of the fan, which is an important part of modern commercial jet engines. Fan increases the total mass flow, thereby enhancing propulsion efficiency. One of the failure modes associated with the fans is flutter vibration due to length of blades and their exposure to inlet flow distortions. It has been known that increasing the length of blades will stiffen the blades, thereby reducing the incidences of flutter vibration. However, increasing the length will cause increase the weight (and cost), which is undesirable. In this invention, as shown in Figure 9.19, blades are made hollow by providing a neck at its inner end that is attached to the hub. This will allow the length of blade to be increased, which will reduce flutter. The strategy used in this invention is *relax a constraint limit*.

Chapter 10

Design for Robustness

Robustness can be defined as designing a product or service in such a way that its performance is the same across all customer usage conditions. Taguchi's methods of designing for robustness are considered powerful and cost effective, and they are aimed at improving the performance of a product or service by reducing the variability across the domain of customer's usage conditions. These methods have received worldwide recognition both in industry and academic community, as they are intended to improve companies' competitive position in the market.

10.1 ENGINEERED QUALITY

Taguchi's approach of design for robustness is based on two classes of quality:

1. Customer quality
2. Engineered quality

Customer quality includes product features such as color, size, appearance, and function. This aspect of quality is directly proportional to the size of the market segment. As customer quality gets better, the market size becomes bigger and companies will have competitive advantage that could result in creation of the new market. Customer quality is addressed during the product planning stage.

The second class, engineered quality, includes defects, failures, noise, vibrations, pollution, and so on. Improving the functionality, most of the time, helps in improving the customer quality. It is important to note that the customer quality defines the market size and the engineered quality helps in winning the market share within the segment. Taguchi methods (TM) of designing for robustness aim at improving the engineered quality.

Most often, engineered quality is not satisfactory because of the presence of three types of uncontrollable or noise factors:

1. Usage conditions (example: environmental conditions)
2. Deterioration and wear (example: degradation over time)
3. Individual difference (example: manufacturing imperfections)

The design of a new product/service for robustness can be achieved in the following three stages:

1. Concept design
2. Parameter design
3. Tolerance design

Most robust design applications focus on parameter design optimization. It is widely acknowledged that the gains in terms of robustness will be greater if you start the designing process with a robust concept selected in the concept stage. Techniques

like P-diagram strategies developed for conceptual robustness can be used to achieve robustness at the concept level. A detailed discussion on this aspect is provided at the end of Chapter 9 of this book (Section 9.4).

The methods of robustness based on Taguchi's approach are developed with the following five principles:

1. Evaluation of the function using energy transformation
2. Studying the interactions between control and noise factors
3. Use of orthogonal arrays and signal-to-noise ratios to improve robustness
4. Two-Step optimization
5. Tolerance design using quality loss function approach

Taguchi (1987) and Phadke (1989) provided a detailed discussion on these topics. Taguchi methods have been successfully applied in many engineering applications to improve the performance of the product/process. They are proved to be extremely useful and cost effective.

10.1.1 Evaluation of the Function Using Energy Transformation

The most important aspect of Taguchi methods (TM) is to find a suitable function (called ideal function) that governs the performance of the system. It helps in understanding the energy transformation in the system by evaluating useful energy and energy spent because of the presence of noise factors. The energy transformation is measured in terms of signal-to-noise (S/N) ratios. Higher S/N ratio means better energy transformation and hence functionality of the system.

10.1.2 Studying the Interactions between Control and Noise Factors

In designing for robustness using Taguchi approach, the interaction between the control and noise factors is exploited since the objective is to make the design robust against the noise factors.

10.1.3 Use of Orthogonal Arrays (OAs) and Signal-to-Noise Ratios to Improve Robustness

Orthogonal arrays (OAs) are used to study various combinations of factors in the presence of noise factors. For each combination, S/N ratios are calculated and used to make decisions about optimal parameter settings. OAs are also helpful to minimize the number of runs (or combinations) needed for the experiment.

10.1.4 Two-step Optimization

After conducting the experiment, the factor-level combination for the optimal design is selected with the help of two-step optimization. In two-step optimization, the first step is to minimize the variability (maximize S/N ratios). In the second step, the sensitivity (mean) is adjusted to the desired level. It is easier to adjust the mean after minimizing the variability.

10.1.5 Tolerance Design Using Quality Loss Function

Although the first four principles are related to parameter design, the fifth one is related to the tolerance design. Having determined the best settings using parameter design, the tolerancing is done to find out allowable ranges to each parameter in the optimal design. This is done using quality loss function, which states that whenever the performance deviates from the

target, there is a loss associated with the deviation. This loss is the *loss to the society*. This loss is proportional to the square of the deviation.

10.2 ADDITIONAL TOPICS IN DESIGNING FOR ROBUSTNESS

In the following sections we provide descriptions of some key topics in design for robustness.

10.2.1 Parameter Diagram (P-diagram)

Parameter or P-diagram is a block diagram, which is often quite helpful to represent a product or a system. It captures all the elements of process just as a cause and effect diagram or SIPOC (suppliers, inputs, processes, outputs and customers) diagram. Figure 10.1 shows all the elements of the P-diagram. The energy transformation takes place between input signal (M) and the output response (y). The goal is to maximize energy transformation by adjusting control factors (C) settings in the presence of noise factors (N).

1. **Signal factors (M)** – These are the factors that are set based on customer usage conditions. These factors play

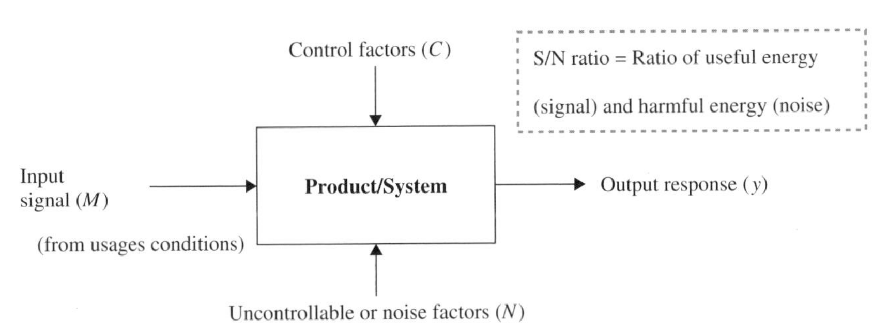

Figure 10.1 Elements of Parameter Diagram or P-diagram.

a significant role in deciding the level of robustness one wants to have. Example: The steering angle is a signal factor for the steering mechanism of an automobile. The signal factors are selected by the engineer based on the engineering knowledge and spectrum of usage conditions they need to consider.

2. **Control factors (C)** – These are the factors that are in the control of the designer. In P-diagram, only control-factor elements can be changed by design engineer by exercising his or her discretion. All other elements are not in the engineer's control, although they play a critical role in robustness. The control factors can take more than one value, which will be referred to as *levels*.
3. **Noise factors (N)** – Noise factors are the uncontrollable factors. The presence of these factors affects the successful energy transformation from the input to output. Since these factors cannot be controlled, it is often desirable to adjust the levels of control factors in such a way that the control factor combination is insensitive to the noise factors, thereby maintaining the same level of performance at all usage conditions. The noise factors can be one of or a combination of the following three types of factors: various usage conditions, deterioration and wear, and individual difference.

10.2.2 Design of Experiments

Design of experiments is a subject that will help investigators to conduct experiments in a systematic fashion and analyze results of experiments by conducting variance analysis to find an optimal parameter combination for design to achieve intended objectives. There is an extensive literature on this subject.

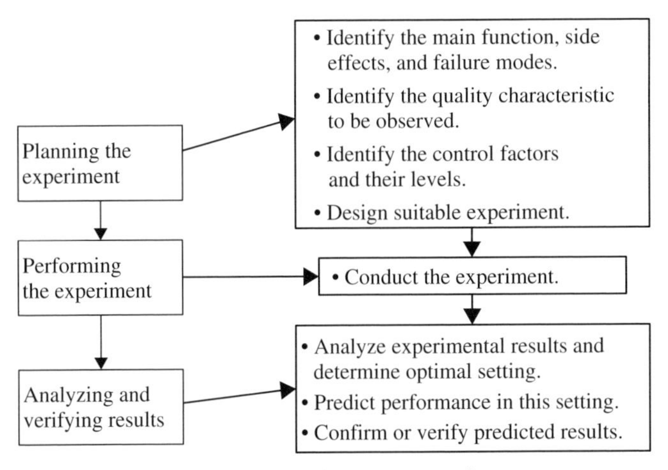

Figure 10.2 Experimental Design Cycle.

A typical experimental design cycle is shown in Figure 10.2, and it consists of the following three important steps:

1. Planning the experiment
2. Performing the experiment
3. Analyzing and verification of experimental results

There are typically two types of experiments:

1. **Full factorial experiments** – In this type, all combinations of factors are studied and the main effects and all possible interaction effects can be estimated using the results of such an experiment.
2. **Fractional factorial experiments** – In fractional factorial experiments, a fraction of the total number of experiments is studied. This is done to reduce cost, material, and time. Main effects and required interactions can be estimated with such experimental results. Orthogonal arrays belong to this class of experiments. Different types of orthogonal arrays are given in Appendix C.

10.2.3 Signal-to-Noise (S/N) Ratios

Signal-to-noise ratio is a metric used determine the magnitude of true output (transmitted from the input signals) after making some adjustment for uncontrollable variation (i.e., noise). The system as shown in Figure 10.1 consists of a set of activities or functions that it needs to perform by producing an intended or desired output by minimizing variations due to noise factors. Usually, the energy transformation in engineering systems takes place through laws of physics.

These engineered systems must be designed to deliver specific results as required by customers. The relationship between the input and the output that governs the energy transformation is referred to as the *ideal function*. When designing for robustness, the ideal function is used as the reference, and the deviation of the actual function from this ideal function is studied (this deviation is proportional to the effect of noise factors). Efforts are made to improve robustness by bringing the actual function close to the ideal. This is shown in Figure 10.3.

If the energy transformation is 100 percent efficient, then there will be no energy loss. As a result, there would be no quality problems or functional failures – that is, no squeaks, rattles, noise, scrap, rework, quality control personnel, customer service agents, complaint departments, or warranty claims. However,

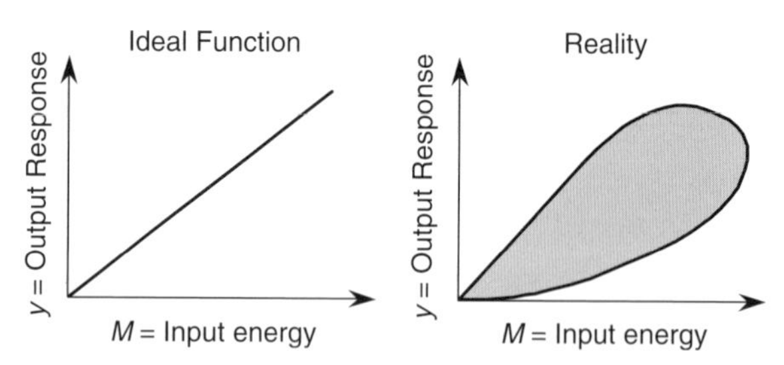

Figure 10.3 Ideal Function and Actual Function.

reality is much different from the ideal situation, and hence, energy transformation cannot be 100 percent. There is going to be energy loss because of noise factors and variation. Higher energy loss indicates the presence of more noise.

The S/N ratio, which is a measure of robustness, is the ratio of energy (or power) that is transformed into intended output to energy (or power) that is transformed into unintended output. We can say that the S/N ratio is the ratio of useful energy and harmful energy. Higher S/N ratio means the system's function is more robust. The equations for different types of S/N ratios are provided in Appendix D.

10.3 ROLE OF SIMULATIONS IN DESIGN FOR ROBUSTNESS

To obtain quick results in the most cost-effective way, simulation experiments are preferred as opposed to hardware experiments. These experiments play a significant role in the design for robustness:

- Simulation experiments play a significant role in reducing product development time because there is no need to conduct all the hardware experiments.
- Simulation-based robust design approach can serve as a key strategy for research and development.
- They help to conduct functionality-based analysis.
- Simulation experiments are typically inexpensive, less time consuming, and more informative, and many control factors can be studied.

After selecting a suitable simulation method, the concept must be optimized for its robustness. The results of simulated experiments are analyzed by calculating S/N ratios and sensitivities, as shown in the following examples. After identifying the optimal design through simulations, it needs to be tested by running a confirmation test. This test will also help us validate the selected model, control factors, and noise factors. In any robust design experiment, it is important to select suitable signals, noise factors, and output response, as shown in the P-diagram.

The example given in Section 10.4 explains how to design for robustness using simulation-based experiments. The approach is the same even if we use hardware experiments or a mix of hardware and simulation-based experiments.

10.4 EXAMPLE – CIRCUIT STABILITY DESIGN

We are thankful to Dr. Genichi Taguchi for allowing us to use this example.

The output current, y (in amperes) of an alternating-current circuit is given by

$$y = \frac{V}{\sqrt{R^2 + (2\pi fL)^2}} \tag{10.1}$$

where V = Input alternating current voltage (V)
R = Resistance (Ω)
f = Frequency of input alternating current (Hz)
L = Self-inductance (H)

and $\omega = 2\pi f$

Let us say that the target value of output current y is 10 amps, with 4 amps of allowable functional range. If y falls outside the functional range, the circuit will not perform its intended function.

While designing for robustness, it is required to determine the optimal parameter levels (normal values and types) of a system of elements. After determining optimal levels, we can use tolerance design to determine the tolerance around the nominal value of the parameters.

10.4.1 Control Factors and Noise Factors

In equation (10.1), only resistance (R) and inductance (L) are control factors. Let us assume that the levels of these factors are as follows:

Resistance R: $R_1 = 0.5\,(\Omega)$	$R_2 = 5.0\,(\Omega)$	$R_3 = 9.5\,(\Omega)$
Inductance L: $L_1 = 0.010\,(\mathrm{H})$	$L_2 = 0.020\,(\mathrm{H})$	$L_3 = 0.030\,(\mathrm{H})$

An orthogonal array is used if there are more control factors. In this case, since there are only two factors (with three levels each), we can select a full factorial experiment for the purpose of experimentation.

Next, let us examine noise factors. As described earlier, noise factors are causes for the deviation of actual function from the ideal function. In this case, dispersion or the deviation can be caused by the two factors' voltage and frequency. The levels that we can consider for these factors are as follows:

Voltage of input source V	90	100	110	(V)
Frequency f	50	60		(Hz)

It is important to note that the variation can also be caused due to the changes the value of the resistance, R, and the coil inductance, L. Because of this, let us decide that the resistance, R, and coil inductance, L, are to have the following three levels:

First level	Normal value $\times$ 0.9
Second level	Nominal value
Third level	Nominal value $\times$ 1.1

The levels of the noise factors are as given in Table 10.1. It should be noted that a prime has been used to denote the noise factors R and L. Here, it is to be noted that there is no error with respect to the noise factors R', and L' when they are at the second level. Frequency f is 50 Hz or 60 Hz, depending on the location (or place of use). If one wishes to develop a product that

Table 10.1 Noise Factors and Levels

	Noise Factor Levels		
	Level 1	*Level 2*	*Level 3*
V	90	100	110 (V)
R'	−10%	0	+10 (%)
f	50	55	60 (Hz)
L$'$	−10%	0	+10 (%)

can be used in both conditions, it is best to design it so that the output meets the target when the frequency is at 55 Hz, midway between the two.

From equation (10.1), the output becomes minimum with $V_1R'_3f_3L'_3$ and maximum with $V_3R'_1f_1L'_1$. If the direction of output changes is known when the noise factor levels are changed, all the noise factors can be compounded into a single factor. If the compound factor is expressed as N, it has two levels. In this case:

$$N_1 = V_1R'_3f_3L'_3 \text{ corresponds to the minimum value}$$

$$N_2 = V_3R'_1f_1L'_1 \text{ corresponds to the maximum value}$$

When we know which noise factor levels cause the output to become large, it is a good strategy to compound them so as to obtain one factor with two levels, or one factor with three levels, including the middle level. When we do this, the noise factor becomes a single compounded factor, irrespective of the factors that are involved. In this example, as we can determine the tendencies of all four noise factors, we have one compounded noise factor with two levels.

10.4.2 Parameter Design

Using equation (10.1), the output current is calculated and design calculations are performed (S/N ratios and sensitivities)

Table 10.2 Calculations of S/N Ratios and Sensitivities

Noise Factors						
No.	*R*	*L*	*Data*		*S/N Ratio*	*Sensitivity*
			N_1	N_2	η	S
1	1	1	21.5	38.5	7.6	29.2
2	1	2	10.8	19.4	7.5	23.2
3	1	3	7.2	13.0	7.4	19.7
4	2	1	13.1	20.7	9.7	24.3
5	2	2	9.0	15.2	8.5	21.4
6	2	3	6.6	11.5	8.0	18.8
7	3	1	8.0	12.2	10.4	20.0
8	3	2	6.8	10.7	9.6	18.6
9	3	3	5.5	9.1	8.9	17.0

for all the nine runs of full factorial experiment. If there are many factors, we would have selected an appropriate orthogonal array in place of full factorial experiment. For each run, data were collected for both levels of compounded noise factor ($N_1 = V_1R'_3f_3L'_3; N_2 = V_3R'_1f_1L'_1$). The assignment of noise factors and experimental results are in Table 10.2.

The S/N ratio is the reciprocal of the square of the relative error (also termed the *coefficient of variation*, $\frac{\sigma}{m}$). Although the equation for the S/N ratio varies, depending on the type of quality characteristic, all S/N ratios have the same properties. When the S/N ratio becomes ten times larger, the loss due to dispersion decreases to one-tenth.

The S/N ratio, η, is a measure for optimum design, and sensitivity, S, is used to select one (sometimes two or more) factor(s) by which to later adjust the mean value of the output to the target value, if necessary.

Table 10.3 Average Factorial Effects (in terms of S/N ratios (η) and Sensitivities (S))

	η	S
R_1	7.5	24.0
R_2	8.7	21.5
R_3	9.6	18.5

	η	S
L_1	9.2	24.5
L_2	8.5	21.1
L_3	8.1	18.5

To compare the levels of control factors, we construct a table of mean values for the S/N ratio, η, and sensitivity, S.

From Table 10.3, we can see that the optimum level of R is R_3, and the optimum level of L is L_1. Therefore, the optimum design is R_3L_1. This combination gives a mean value of 10.1, which is very close to the target value. If there is no difference between the mean value and the target value, we might consider this as optimal design and, therefore, we don't have to adjust control factors based on sensitivities. If there is a difference, it is required to compare the influence of the factors on the S/N ratio and on the sensitivity, and use a control factor or two whose effect on sensitivity is high compared with its effect on the S/N ratio in order to adjust the output to the target.

10.5 PCB DRILLED-HOLE QUALITY IMPROVEMENT

We are thankful to Mr. R. C. Sarangi and Prof. A. K. Choudhury for their help in this work.

This case study was conducted in a well-known PCB industry. This study is related to a drilling operation in PCB manufacturing. This simple study shows how the Taguchi approach and metrics such as S/N ratios and sensitivities are applicable in the real world.

Drilling is one of the most important operations of printed circuit board (PCB) manufacturing. There are several characteristics that determine drilled-hole quality. All these characteristics were usually treated as the qualitative factors and inspected for their presence or absence through visualization. In this study, a methodology by which hole quality could be expressed in quantitative terms was identified. Using parameter design approach, hole quality was improved from 0.1 to 6.4.

10.5.1 Introduction

A printed circuit board (PCB) is the base electronic component with electrical interconnections on which several components are mounted in a compact manner to give the desired electrical output. Drilling is one of the most important operations of PCB manufacturing, since drilled hole is considered as the heart of a PCB. Drilling is carried out on the panels. A panel consists of one or more circuits, which will be routed to the required shape during the routing operation. Usually a stack consisting of more than two panels is used for drilling. The number of panels in a stack is referred to as *stack height*. The drilled holes are meant for the following.

- Producing an opening through the board to permit subsequent processes to form electrical connections between different layers (in case of double-side PCBs)
- Permitting through the board, the component mounting with structural integrity and precision of location

10.5.2 Drilled-hole Quality Characteristics

Drilling defects can be classified as copper defects and substrate or Epoxy defects. The defects under each category are as shown in Table 10.4

The definitions of these defects are presented in Table 10.5 and Table 10.6. All these defects are usually treated as the qualitative factors and inspected for their presence or absence through visualization.

10.5.3 Background

There were several customer complaints regarding poor hole quality. Most of these complaints were from overseas customers. Therefore, it became the need of the hour for this PCB manufacturing company to study drilling process with a dedicated team.

The team started with a reference study to identify a methodology by which hole quality could be expressed in quantitative terms. After doing an extensive survey, a methodology was identified [Coombs, 1988]. This method, called *hole-quality standard*, was essential since quality improvement could not be carried out without a quantitative measurement.

Table 10.4 Drilling Defects

Copper Defects	*Substrate Defects*
Delamination	Delamination
Nail heading	Voids
Smear	Smear
Burr	Plowing (roughness)
Debris	Debris pack
Roughness	Loose fibers

Table 10.5 Measuring Hole Quality (Copper Defects)

Defects	*Definition*	*Weightage Factor* (a_i)	*Extent Factor* (b_i)
Burr (A ridge on the outside of the copper surface after drilling)	Burr height (microns)		
	1.524		0.01
	4.864	1.0	0.08
	7.874		0.30
	12.954		1.20
Nail heading (A flared condition of internal conductor)	Nail head width (microns)		
	3.040		0.01
	5.440	1.5	0.08
	15.740		0.30
	25.900		1.20
Smear (Fused deposit on the copper from the excessive drilling heat)	Percentage copper area covered with smear		
	1%		0.01
	11%	1.5	0.08
	26%		0.30
	36%		1.20

10.5.4 Hole-quality Standard

The bottom panel was always used for the analysis of hole quality. *Coupon holes* were designed on the panel in such a way that four holes were drilled for each hit region of interest. One hit was equivalent to drilling one hole. This is described pictorially in the Figure 10.4.

Table 10.6 Measuring Hole Quality (Substrate Defects)

Defects	*Definition*	*Weightage Factor (a_i)*	*Extent Factor (b_i)*
Voids (A cavity in the substrate)	Minimum magnification required to see the defects clearly		
	140X (*)	0.8	0.01
	100X		0.08
	60X		0.30
	20X		1.20
Debris pack (Debris deposited in cavities)	Percentage substrate area covered with debris		
	1%	0.8	0.01
	11%		0.08
	26%		0.30
	36%		1.20
Loose fibers (Supporting fibers in the substrate of a laminate that are not held in	Percentage substrate area covered with loose fibers	0.3	0.01
	1%		
place by	11%		0.08
surroundings	26%		0.30
resin)	36%		1.20
Smear (Fused deposit left on the substrate from	Percentage substrate area covered with smear		
the excessive	1%		0.01
drilling heat)	11%	0.3	0.08
	26%		0.30
	36%		1.20
Plowing (Furrows in hole wall due to drilling)	Minimum magnification required to see the defect clearly		
	140X	0.2	0.01
	100X		0.08
	60X		0.30
	20X		1.20

(*)times magnification using a microscope

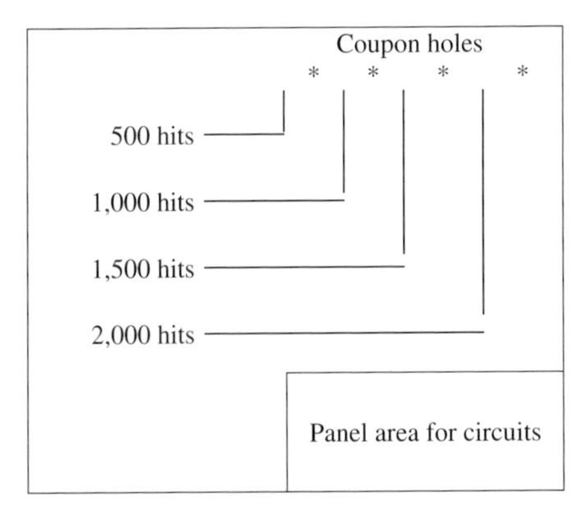

Figure 10.4 Coupon Holes on a Panel.

The coupon holes are removed from the bottom laminate by using a cutting wheel. The coupons are then molded by thermosetting plastic material. The holes in the mold are sanded down (grinding and polishing) until holes are opened to its diameter. In this manner not much of the hole is wasted and most of the hole wall could be examined through a microscope. Preparation of the mold and its operations would take about two hours. The coupon holes are then seen through the microscope under different magnifications, as defined in Table 10.5 and Table 10.6.

Hole quality can be determined by using the following formula.

$$\text{Hole quality} = 10\,(0.2)^{\Sigma a_i b_i} \tag{10.2}$$

where $\Sigma a_i b_i$ = Sum of products weightage factor and extent factors of all defects.

The extent factor corresponding to a value, which is not given in Tables 10.5 and 10.6, is determined by the linear interpolation.

Hole quality lies between 0 and 10. Any value above 6.0 is considered satisfactory. If the value was above 7.0, then it is proposed that the panels need not undergo subsequent operations such as desmearing and deburring, resulting in reduction of production time.

Table 10.7 Existing Hole Quality

Copper Defects	*Value*	a_i	b_i	$a_i\, b_i$
Burr height	17.5 microns	1.0	2.0	2.0
Nail head width	10.0 microns	1.5	0.245	0.367
Smear	21.42%	1.5	0.267	0.400
Substrate defects				
Voids	100X	0.8	0.08	0.064
Plowing	100X	0.2	0.08	0.016
Smear	5.46%	0.3	0.04	0.012
Loose fibers	0.468%	0.3	0.005	0.0015
Total				**2.8605**

The first task of the team was to estimate the existing hole quality. The details of the existing hole quality are given in Table 10.7.

$$\text{Hole quality} = 10\,(0.2^{2.8605}) = 0.100$$

Since the existing hole quality was very low, it was decided to improve the level of hole quality through the parameter design approach.

10.5.5 Experiment Description

After several discussions, the following factors were considered to have greater influence on overall drilling quality.

1. Cutting data (feed, speed)
2. Number of hits
3. Drill bit type
4. Product hole-quality positioning
5. Panels, copper foil, drill size, and drill depth
6. Drill aides (entry and back up material)

10.5.6 Selection of Levels for These Factors

For a given drill diameter, spindle speed, retraction speed, surface feet per minute (SFM), and range of feed are fixed. Since four layer boards and 0.95 mm diameter drill bits are commonly used, they were considered for the purpose of this study. For 0.95 mm, hole diameter fixed factors are as shown in Table 10.8.

The factors and levels, as shown in Table 10.9, were considered for the purpose of experimentation.

In Table 10.9, factor B was the supporting material for the panels to facilitate drilling operation. Factor C is the number of panels drilled at a time. Factors D is number of holes drilled per drill bit. Neck-relieved drill bits, also known as undercut

Table 10.8 Fixed Factors for 0.95 mm Drill Diameter

Factor	*Fixed Level*
Feed	121 to 156 IPM (^)
Spindle speed	51 KRPM (^^)
Retraction speed	500 IPM
SFM	500

(^)Inch per minute;
(^^)Kilo revolutions per minute

Table 10.9 Factors and Levels

Factor	*Level 1*	*Level 2*
Feed (A)	121 IPM	138 IPM
Entry and back-up material (B)	LCOA (Aluminum)	HYLAM (Phenolic)
Stack height (C)	3 High	4 High
Number of hits (D)	2,000	1,500
Drill bit type (E)	Ordinary (OD)	Neck relieved (NR)

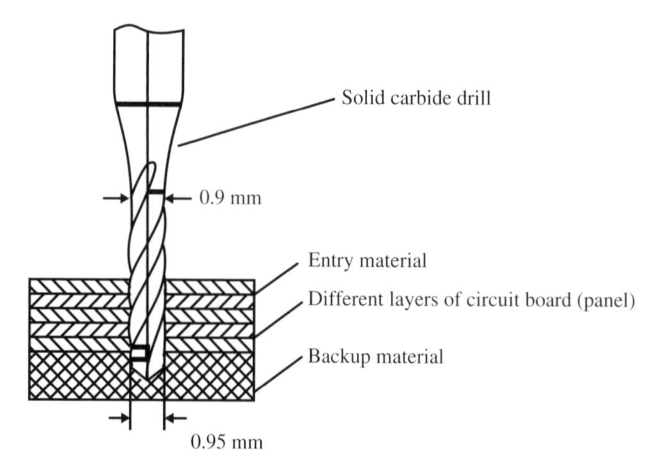

Figure 10.5 Undercut Drill Bit.

drill bits (see Figure 10.5), have an effective diameter up to 0.8 mm. Because of this, while drilling the panels, at a time, the contact time between the hole wall and the drill bit is less. The enormous amount of the heat generated (because of 51 KRPM) will not get transferred to the hole wall. Because of this, factor E was considered.

Since drilling operations are carried out in a controlled environment, external noise factors were not considered in this experiment. The noise that was considered was inner noise due to manufacturing variations.

10.5.7 Designing the Experiment

Since all main effects and interactions AD and CD were considered to be important $L_8(2^7)$, orthogonal array was chosen for the experiment. The factors were allocated to the columns of $L_8(2^7)$ orthogonal array with the help of linear graphing. The factor allocation is shown in Table 10.10.

The physical layout of the experiment is given in Table 10.11.

For each test, hole quality was measured four times. Since our goal was to maximize the hole quality, the difference between the

Table 10.10 Factor Allocation in $L_8(2^7)$ Orthogonal Array

Test \ Column	1 (D)	2 (A)	3	4 (B)	5 (E)	6	7 (C)
1	1	1	1	1	1	1	1
2	1	1	1	2	2	2	2
3	1	2	2	1	1	2	2
4	1	2	2	2	2	1	1
5	2	1	2	1	2	1	2
6	2	1	2	2	1	2	1
7	2	2	1	1	2	2	1
8	2	2	1	2	1	1	2

Table 10.11 Physical Layout of the Experiment

Test	A	B	C	D	E
1	121 IPM	LCOA	3 High	2,000 hits	OD
2	121 IPM	HYLAM	4 High	2,000 hits	NR
3	138 IPM	LCOA	4 High	2,000 hits	OD
4	138 IPM	HYLAM	3 High	2,000 hits	NR
5	121 IPM	LCOA	4 High	1,500 hits	NR
6	121 IPM	HYLAM	3 High	1,500 hits	OD
7	138 IPM	LCOA	3 High	1,500 hits	NR
8	138 IPM	HYLAM	4 High	1,500 hits	OD

observed hole quality and the target value (10) was considered as response. Accordingly, smaller-the-better type S/N ratios were considered for the analysis. The S/N ratio, η, for each test was calculated using the following formula:

$$\eta = -10 \log (1/n)[\Sigma(X_j - P)^2] \tag{10.3}$$

where n = number of observations in each test

$X_j = j^{th}$ value of observed hole quality in a test; $j = 1,2,3,4$

P = Maximum value of hole quality = 10.0

The S/N ratios (in dB) of these tests are given in Table 10.12.

The results of the experiment were analyzed based on S/N ratios. The details of the analysis are shown in Table 10.13. Average response curves for S/N ratios were as shown in Figure 10.6.

Table 10.12 Results of the Experiment

Test	*S/N Ratio (dB)*
1	−19.293
2	−15.119
3	−19.964
4	−14.378
5	−15.544
6	−18.868
7	−11.375
8	−19.330

Table 10.13 Analysis with S/N Ratio as the Response

Source	*Degrees of Freedom*	*Sum of Squares*	*Mean Squares*	*F Ratio*	*Contribution Ratio (%)*
A	1	1.859	1.859	1.323	0.684
B	1	1.619	1.619	1.152	0.322
C	1	4.686	4.686	3.335	4.95
D	1	1.727	1.727	1.229	0.485
E	1	54.909	54.909	39.081	80.687
Error	2	2.81	1.405		12.872
Total	7	67.61			

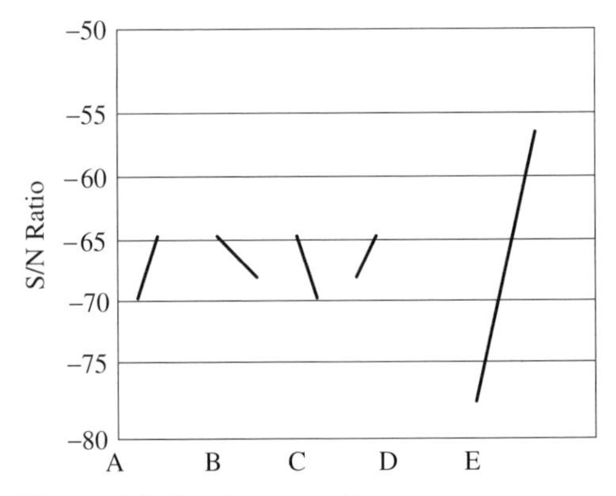

Figure 10.6 Average Responses.

From this analysis, the optimal combination was found to be as follows:

Feed (A2)	: 138 IPM
Entry and Back-up (B1)	: LCOA
Stack height (C1)	: 3 High
Number of hits (D2)	: 1500
Drill bit type (E2)	: Undercut

10.5.8 Predictions and Confirmation Run

For the optimal combination, the predicted S/N ratio was estimated as follows:

$$\eta = (A2 - \tau) + (B1 - \tau) + (C1 - \tau) + (D2 - \tau) + (E2 - \tau) + \tau \tag{10.4}$$

where τ = Overall average of all S/N ratios = −66.98 dB. The other terms on right hand side of the equation represents average S/N ratios corresponding to A2, B1, C1, and D2, and hence $\eta = -47.83$ dB. The corresponding hole quality was 6.95.

A Confirmation run was performed for the optimal combination, and it gave us an S/N ratio of −52.39 dB. The corresponding hole quality was 6.4.

Encouraged by these results, the optimal combination has been implemented in actual production. The hole quality before and after the experiment was shown in Figure 10.7 and Figure 10.8, respectively.

10.5.9 Benefits

The primary benefit associated with this study was increased customer satisfaction. Overseas customers were very much

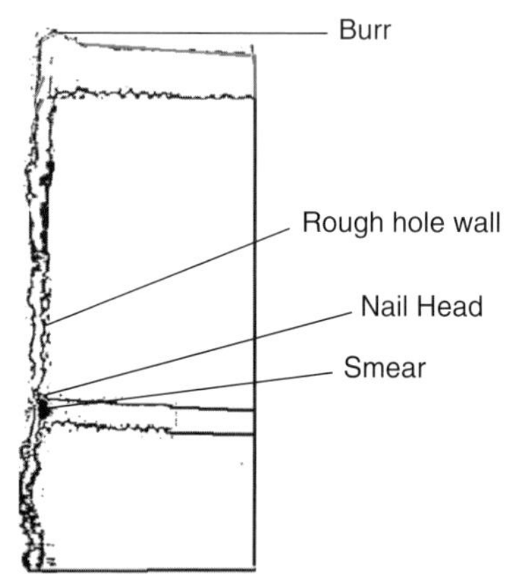

Figure 10.7 Hole Quality 0.1 (Before Experiment).

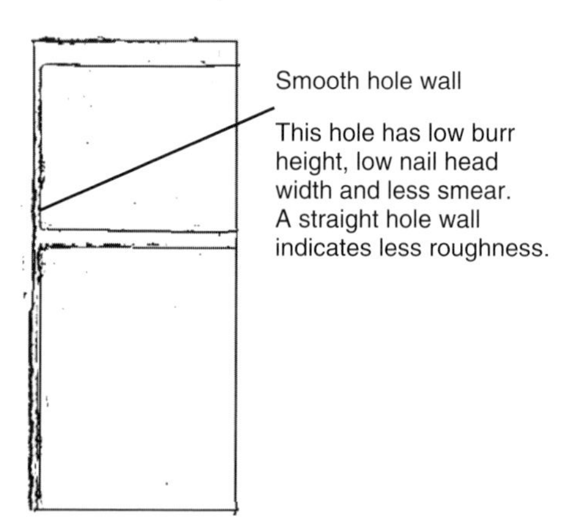

Figure 10.8 Hole Quality 6.4 (After Experiment).

appreciative of the study. The number of orders increased significantly after this study. Better hole quality resulted in better solderability, which attracted more customers, thereby increasing the company's global market share.

10.6 DESIGN OF A VALVELESS MICROPUMP USING TAGUCHI METHODS

We are thankful to Dr. Il Yong Kim of Queen's University, Canada, and Dr. Akira Tezuka of National Institute of Advanced Industrial Science and Technology, Japan, for allowing us to publish this case study.

This example presents robust design of a bidirectional valveless pump. In the previous research, behaviors of the micropump using multiphysics analysis and design optimization were studied. The deterministic analysis that assumes that parameters are exact in their values ignores uncertainties in fabrication and control. Therefore, the optimal design may not function well under uncertainty. Here, six noise factors, along with three control factors, are used for robust design of the pump. A parametric finite element analysis model is made available for the estimation of the micropump performance for the control factors and noise factors.

10.6.1 Introduction

Micropumps are devices that are used to transfer small amount of liquid. In order to achieve rectification, which is an essential function in pumping, most micropumps utilize mechanical valves. These valves are prone to wear and fatigue, and miniaturizing devices is often difficult due to the mechanical valves. Dynamic valve pumps or valveless pumps, which do not use moving valves, were developed as an alternative [Stemme et al., 1993; Olsson et al., 1996; Gerlach et al, 1996; Gerlach et al., 1996; Matsumoto et al., 1999].

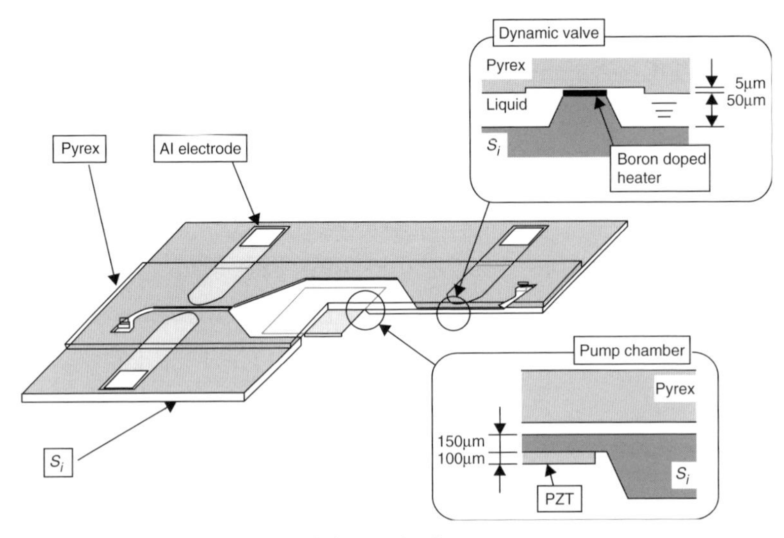

Figure 10.9 Structure of the Valveless Micropump.
Source: From Matsumoto et al. (1999) [© 1999 IEEE].

The structure of the micropump developed by Mastumoto appears in Figure 10.9. Some performance indexes, e.g. total volume flow rate, are relatively easy to estimate experimentally, but numerical simulation is often needed to see some other behaviors. For example, temperature distribution and velocity profile in the device are important parameters, but physical testing can hardly estimate them.

In an earlier study [Kim and Tezuka 2003], a Finite Element Analysis model for design optimization was made. The entire domain of the micropump was modeled, and transient multiphysics numerical analyses were conducted for the whole working cycle. Two important performance metrics – volume flow rate and rectification efficiency – were used as objective functions for design optimization. Heat flux, preheating time, and heating time were designated as design variables because these electrical input signals can be controlled easily by operators.

In the previous research, all parameters were assumed to be exact in their values, and uncertainties in fabrication and control were not considered. However, fabrication errors are often large, especially in MEMS (micro-electro-mechanical systems) applications. These errors, along with control errors, could affect the system's performance significantly. Taguchi method was applied to determine a robust design. The three design variables were used as control factors. Three noise factors in both control and fabrication were considered for robust design.

10.6.2 Working Principle and Finite Element Modeling

Figure 10.10 shows the structure of the micropump. The thicknesses of the Pyrex substrate and the silicon substrate are 500 μm, and the height of the fluid chamber is 50 μm. Two narrow channels of 5 μm height function as valves for rectification. Electric heaters are placed on the channels, and the piezoelectric plate actuates the central diaphragm of the pump chamber.

The working cycle is composed of two modes – a pumping mode and a supplying mode. In a pumping mode, electricity is applied to the PZT, and the diaphragm goes up, squeezing out the working liquid through the two channels. At this time, the

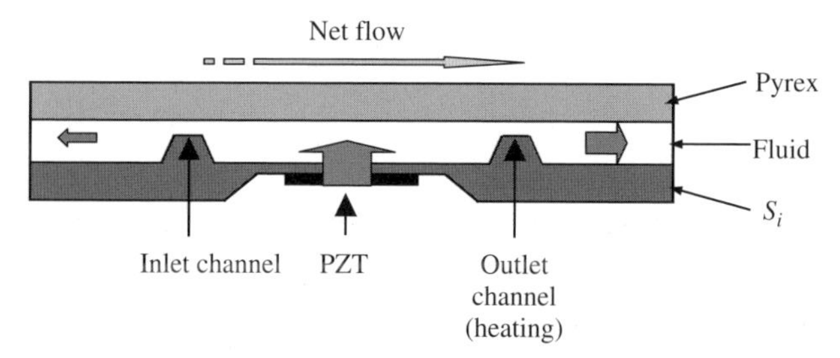

Figure 10.10 Basic Structure and Working Principle (Pumping Mode is Shown).

right-side channel is heated, which results in a viscosity drop in the region, but the left-side channel is not affected by this heating. The unsymmetrical viscosity produces asymmetrical flow resistance in two channels, and more liquid flows out through the right-side channel.

In a supplying mode, the central diaphragm goes down to its original position, sucking the working liquid in through two channels. In this mode, the left-side channel is heated, and the right-side channel is cooled down. More liquid flows in through the left-side channel because of lower viscosity in the region. The left-side and right-side channels are called Inlet and Outlet channels, respectively. These two working modes are repeated alternately and rapidly producing a net flow of the liquid in the rightward direction. A net flow in the opposite direction can be made if the left-side channel is heated during the squeezing, and the right-side channel is heated during the suction. The ability to perform bidirectional pumping is a great advantage of this type of pump.

A short preheating is conducted before the diaphragm movement in each mode. This preheating enables more efficient rectification because the viscosity will have become low enough when the flow begins. The analysis domains, boundary conditions, and loads are shown in Figure 10.11. The three parameters – heat flux magnitude, preheating time, and heating time – are controlled electrically and determine the behaviors of the liquid flow for a given movement of the PZT.

10.6.3 Design for Robustness

The goal of robust design in this work is to achieve maximum and accurate volume flow rate in the presence of noise factors. In the computational design optimization of previous research [Kim 2003], the volume flow rate was maximized based on the assumption that all dimensions and signal controls are exact.

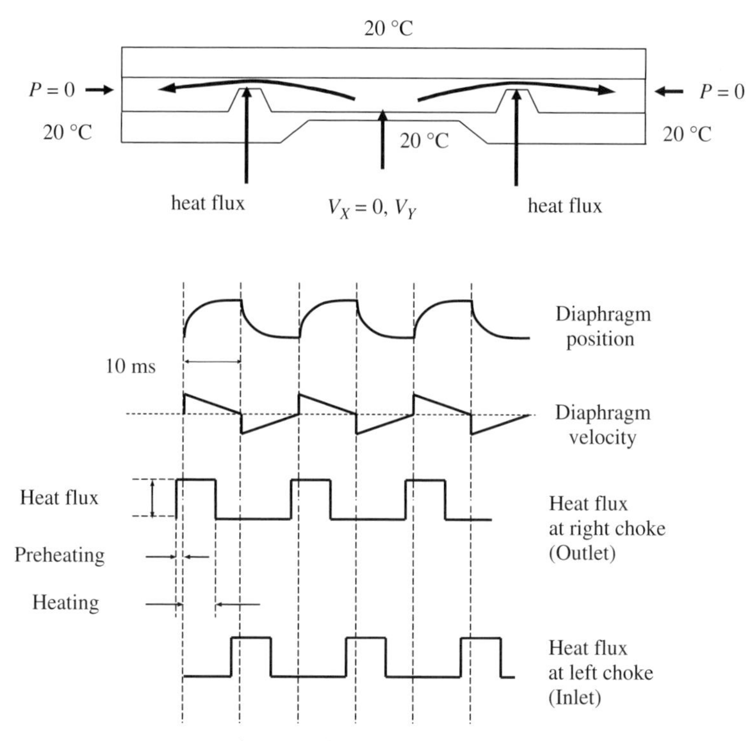

Figure 10.11 Loading and Boundary Conditions.

In real situations, however, these parameters have uncertainties, and the optimum design that is obtained deterministically would not perform as intended. A simulation-based robust engineering method [Taguchi, Jugulum, Taguchi 2004] was used to determine an optimum design that is robust to these uncertainties.

Control factors, which designers change to control the output of the system, are the same as the design variables in the previous optimization. Six noise factors – errors or variation in the control factor settings – and three fabrication errors are considered. Unlike macro manufacturing, the dimensional uncertainties associated with MEMS fabrication are large, and in this case in particular, the PZT positioning, which is done

Table 10.14 Control Factors and Noise Factors

Control Factors

		L1	*L2*	*L3*
A	Heat flux	50	75	100
B	Preheating time	0.1	0.5	0.9
C	Heating time	1.0	5.0	9.0

Noise Factors

		N1	*N2*
A	Heat flux	−5 %	+5 %
B	Preheating time	−5 %	+5 %
C	Heating time	−5 %	+5 %
D	Left gap size	4.9	5.1
E	Right gap size	4.9	5.1
F	PZT position	1450	1550

manually, has great uncertainty. Table 10.14 shows the control and noise factors. Note that the full combinatorial cases of the noise factors are not considered, but the factors are compounded in this study.

Experimentation is conducted according to L_9 orthogonal array experimentation table. As shown in Table 10.14, there are six noise factors. We have used compounding of noise strategy to reduce the number of experiments. By using compounding strategy, we have only one compounded noise factor with two levels, N_1 and N_2. The levels N_1 and N_2 are obtained as follows: For N_1: the factors A, B, C are set −5 percent of the chosen levels (in orthogonal array experiment), and the remaining factors are set at the levels shown in Table 10.14. Similarly, for N_2: the factors A, B, C are set 5 percent of the chosen levels (in orthogonal array experiment), and the remaining factors are set at the levels shown in Table 10.14. With this set-up, experiments

Table 10.15 Experimental Results

Experiment	*Control Factor*			*Noise*	*Objective*	*Constraint*
No.	A	B	C	*Factor*	*Function* ($\times 10^{-11}$)	*Temp*<*80*
1	L_1	L_1	L_1	N_1	0.3248	41.9
2	L_1	L_1	L_1	N_2	0.3607	43.2
3	L_1	L_2	L_2	N_1	1.3748	44.8
4	L_1	L_2	L_2	N_2	1.4944	46.3
5	L_1	L_3	L_3	N_1	1.8215	46.4
6	L_1	L_3	L_3	N_2	1.9126	48.1
7	L_2	L_1	L_2	N_1	1.8540	57.1
8	L_2	L_1	L_2	N_2	2.0077	59.3
9	L_2	L_2	L_3	N_1	2.4828	59.6
10	L_2	L_2	L_3	N_2	2.5927	62.0
11	L_2	L_3	L_1	N_1	0.5530	54.7
12	L_2	L_3	L_1	N_2	0.6057	56.7
13	L_3	L_1	L_3	N_1	2.9893	72.7
14	L_3	L_1	L_3	N_2	3.1055	76.0
15	L_3	L_2	L_1	N_1	0.6577	65.1
16	L_3	L_2	L_1	N_2	0.7209	67.8
17	L_3	L_3	L_2	N_1	2.3130	69.5
18	L_3	L_3	L_2	N_2	2.4848	72.3

were conducted by using L_9 orthogonal array with two levels of compounded noise factor. Table 10.15 shows the results of the experiment, and the measure of objective function in this table is volume flow rate. Table 10.16 shows the results of data analysis. In robust engineering approach [Taguchi, Jugulum, Taguchi, 2004], we usually calculate signal-to-noise ratios (S/N Ratio) and sensitivities for the purpose of optimization.

After computing these two quantities for all experimental combinations of the orthogonal array, two-step optimization is performed to identify the optimal design. This is a very important optimization strategy in robust engineering. In the first step, we seek a design that maximizes the signal-to-noise ratio or minimizes variability. In the second step, we will adjust the response to meet the requirement. Then the design is validated only if it meets all other requirements. Variability is the key issue in robust design. It is usually difficult to reduce variability as compared to adjusting the process to meet requirements. This is the reason that we should aim at reducing variability first.

Depending on the type of design problem, we use an appropriate S/N ratio. Typically, there are five types of S/N ratios, depending on the type of the response:

1. **Smaller the better type** – The response is continuous and positive. Its most desired value is zero. Examples are the offset voltage of a differential operational amplifier, the pollution from a power plant, and the leakage current in integrated circuits.
2. **Nominal-the-best** – The response is continuous and it has a nonextreme target response. For example, we would want all contact windows to have the target dimension, say 3 microns, in the window photolithography process of integrated circuit fabrication.
3. **Larger the better** – This is the case of continuous response where we would like the response to be as large as possible. The strength of a material is an example of this class of problems.
4. **Ordered categorical response** – This is the case where the response is categorized into ordered categories such as very bad, bad, acceptable, good, very good.
5. **Dynamic problems** – Dynamic problems are characterized by the presence of signal factor (M) and the response variable (y). The signal factor corresponds to different customer usage conditions. The robust design is obtained by finding factor settings such that output y is minimally affected by noise factors. Among these S/N ratios, dynamic S/N ratio is considered to be most important because we

can optimize the product by testing it under all customer usage conditions.

In this application we have used a dynamic type of S/N ratio. The signal factor is the flow rate at nominal or standard condition. Table 10.16 shows dynamic S/N ratios (SNR) and sensitivities (S) for all combinations of orthogonal array. The equations for computing these quantities in decibel (dB) units are as follows:

$$SNR = 10\text{Log}\left[\frac{1/n(Sm - Ve)}{Ve}\right]$$

$$S = 10\text{Log}[1/n(Sm - Ve)]$$

where Sm is the sum of squares due to mean; Ve is the error variance, and n is the number of observations (responses) in each combination.

Based on the information in Table 10.16, average responses with respect to S/N ratios and sensitivities are computed for three levels. Figure 10.12 and Figure 10.13 summarize this information.

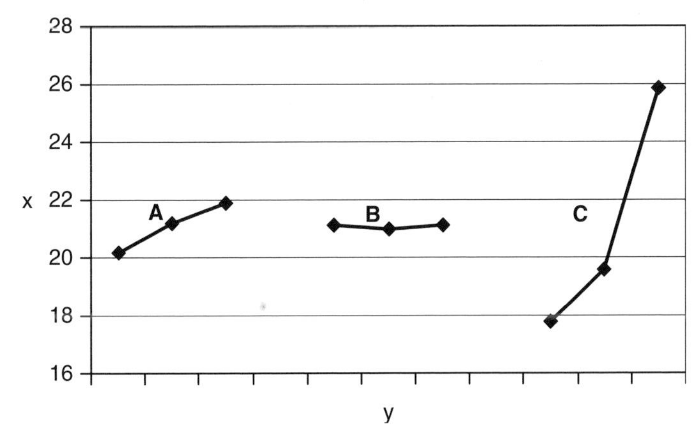

Figure 10.12 S/N Ratio Analysis.

Table 10.16 Data Analysis

Experiment	*Control Factor*			N_0	N_1	N_2	*Mean*	*Sd*	*Ve*	*Sm*	*Sm−Ve*	*1/n(Sm−Ve)*	*SNR*	*S*
No.	*A*	*B*	*C*											
1	L1	L1	L1	0.3605	0.3243	0.3972	0.3608	0.0515	0.0027	0.2603	0.2576	0.1288	16.8553	−8.9004
2	L1	L2	L2	1.4928	1.3743	1.6106	1.4925	0.1671	0.0279	4.4548	4.4269	2.2134	18.9917	3.4507
3	L1	L3	L3	1.9087	1.8215	1.9788	1.9002	0.1112	0.0124	7.2211	7.2088	3.6044	24.6440	5.5683
4	L2	L1	L2	2.0067	1.8540	2.1569	2.0055	0.2142	0.0459	8.0437	7.9978	3.9989	19.4037	6.0194
5	L2	L2	L3	2.5886	2.4828	2.6696	2.5762	0.1321	0.0174	13.2736	13.2562	6.6281	25.7966	8.2139
6	L2	L3	L1	0.6041	0.5530	0.6564	0.6047	0.0731	0.0053	0.7313	0.7260	0.3630	18.3188	−4.4011
7	L3	L1	L3	3.1020	2.9893	3.1830	3.0862	0.1370	0.0188	19.0486	19.0299	9.5149	27.0518	9.7841
8	L3	L2	L1	0.7196	0.6577	0.7830	0.7204	0.0886	0.0079	1.0378	1.0300	0.5150	18.1692	−2.8821
9	L3	L3	L2	2.4826	2.3130	2.6477	2.4804	0.2367	0.0560	12.3043	12.2483	6.1241	20.3876	7.8704

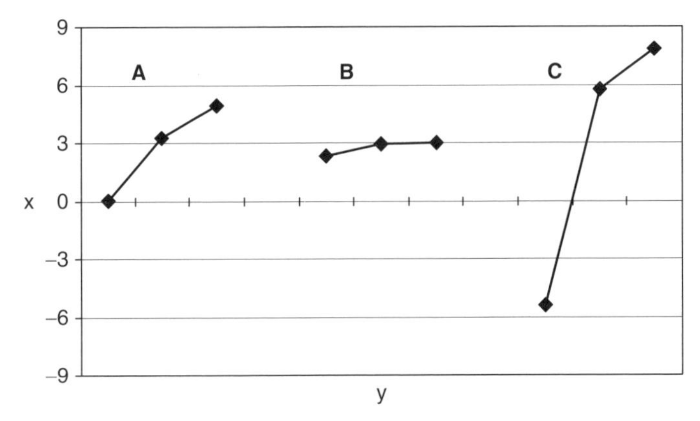

Figure 10.13 Sensitivity Analysis.

Based on these averages and sensitivities, two-step optimization is performed as follows.

Step 1

In this step, the combination that maximizes the S/N ratio is identified. This combination is found to be A3-B1-C3. At these respective levels of factors, the S/N ratios are highest. This combination is part of L_9 array (combination #7). For this combination, the value of SNR is 27.05 dB units and sensitivity is 9.78 dB units. The corresponding mean flow rate is 3.09×10^{-11} m^3/s.

Step 2

Since the target flow rate 3.30×10^{-11} m^3/s [Kim and Tezuka 2003], it is necessary to adjust the mean to target. If we look at Figure 10.12, factor B has the least impact for S/N ratios. Therefore, this factor can be considered for adjusting the mean to target. Looking the Figure 10.13, it is clear that factor B has the higher sensitivity at level 3. If we change the level of factor B to B3 from B1 and keep factors A and C at the same levels (A3 and C3), the corresponding predicted S/N ratio and sensitivity would be SNR $= 26.66$ dB units and $S = 10.21$ dB units. The corresponding predicted flow rate would be 3.24×10^{-11} m^3/s. Hence, by changing the level of B to B3, we can obtain the desired flow rate, although we will have to compromise a little bit on SNR. Therefore, the optimal combination is A3-B3-C3. When we performed confirmation run for this combination, surprisingly it gave us an S/N ratio of 31.71 dB units with flow rate of 3.09×10^{-11} m^3/s. This value of S/N ratio is the highest we observed. It was decided to use this combination because of higher S/N ratio (more robust), although the flow rate is somewhat lower.

10.6.4 Conclusions

Robust design of a valveless micropump is conducted using the Taguchi methods. In the previous study, all electrical input signals and dimensions were assumed to be accurate in their values. In this study, noise factors for the three input signals (heat flux, preheating time, and heating time) and three critical dimensions (left gap size, right gap size, and PZT position) were considered for robust design. Dynamic S/N ratios and sensitivities were computed (See Table 10.16), and two-step optimization was conducted. In the first step, where the variation is minimized, A3-B1-C3 is the optimal design. However, the optimal combination in the second step is A3-B3-C3, which maximizes volume flow rate and S/N ratio.

Chapter 11

Robust System Testing

This chapter describes a methodology that can be used to test the performance of a given system after designing the same. This procedure uses the principles of robust engineering, in particular two-level orthogonal arrays. This procedure is described with the help of successful case applications. This method uses orthogonal arrays (OAs) to study the effect of two factor (active signal) combinations. Usually, it is sufficient to study two factor combinations because, higher order effects are small and hence they can be neglected.

11.1 INTRODUCTION

Any given system should perform its intended function under all combinations of the user conditions. These user conditions are referred to as active signals and they are used to get the desired output. Examples of the active signals are: inserting the card, punching the personal identification number (PIN), and applying pressure on brake. For any given system the user conditions are unique, and they can be numerous. Usually, the designers test the performance under the user conditions separately (one factor at a time). Even after such tests, the system fail because of the presence of interactions between the active signals. Therefore, the designer must study all the interactions and take appropriate corrective actions before the release of the product. The presence of interactions can be obtained by studying two-factor combination effects. To obtain these effects, the software should be tested under various combinations of the signals.

The different states of the signal are referred to as the different *levels*. For a given signal, the number of levels may be very high. In the case of ATM transaction example, the levels for PIN may be 0001 to 9999. If the number of such signals is very high, then the number of possible combinations will be in the billions. Since it is not feasible to test the system under all the combinations, a procedure is necessary to minimize the number of combinations. In this chapter, such a method is developed by using the principles of robust engineering. Using this procedure, all possible two-factor combination effects can be obtained by conducting almost the same number of experiments as in the case of one-factor-at-a-time experiments.

11.1.1 A Typical System Used in Testing

A typical system used for testing is shown in Figure 11.1. This is similar to a p-diagram showing the typical testing arrangement used this approach. In this figure, the noise factors correspond to the hardware conditions for the software example. For the ATM example, the noise factor is the condition of the ATM machine (old or new). Control factors refer to system specifications. Here, we are basically interested in user conditions, as they are primarily used to develop test procedures with the help of orthogonal arrays.

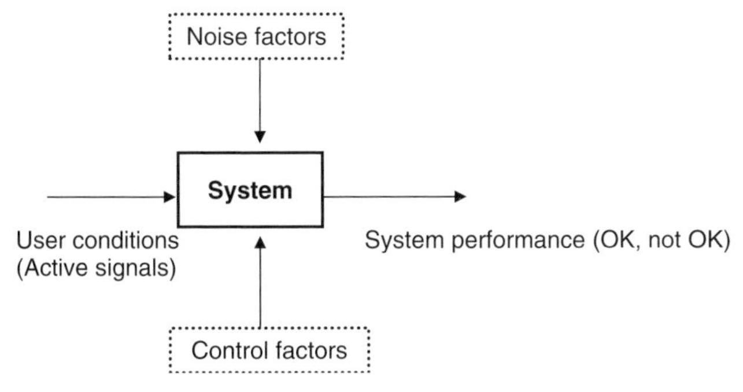

Figure 11.1 P-diagram for Software Testing.

11.1.2 Role of the Orthogonal Arrays

The purpose of using the orthogonal arrays in the design for robustness is to estimate the effects of several factors and required interactions by minimizing the number of experiments. In the case of testing, the purpose using the OAs is to study all the two-factor combinations with a minimum number of experiments. As mentioned earlier, the number of combinations with OAs is equal to the number of experiments to be conducted with one factor at a time.

Let us suppose that there are twenty-three active signals, eleven with two levels and the remaining twelve with three levels. If we want to study all the two-factor combinations, the total number of experiments in this case would be 1,606. For this example, if a suitable OA is used, the number of experimental runs to obtain all two-factor combinations would be thirty-six. The corresponding array is L_{36} $(2^{11} \times 3^{12})$ where L denotes the Latin square design; 36 is the number of test runs; 11 is the number of two-level signal factors that can be used, and 12 is the number of three-level signal factors that can be used. Please refer to Appendix C for listing of orthogonal arrays.

11.2 METHOD OF SOFTWARE TESTING

For the purpose of explanation, let us consider the same example with twenty-three signals. Let A,B,C,..,L,M,..,U,V,W represent these twenty-three factors. The signal factors are allocated to the different columns of the L_{36} array, as shown in Table 11.1. In the array the numbers 1,2,3 correspond to the different levels of the signals. The 36 test runs in Table 1 have to be performed to obtain the effect of two factor combinations. The response for each combination is 0 or 1. 0 means the software function

Table 11.1 Signal Factor Allocation in L_{36} ($2^{11} \times 3^{12}$) Array

Signal Factor	*A*	*B*	.....	*L*	*M*	.....	*V*	*W*	
Run / Column	*1*	*2*	.....	*12*	*13*	.....	*22*	*23*	*Response*
1	1	1		1	1	.....	1	1	0 or 1
2	1	1		2	2	.....	2	2	0 or 1
3	1	1		3	3	.....	3	3	0 or 1
4	1	1		1	1	.....	3	3	0 or 1
5	1	1		2	2	.....	1	1	0 or 1
6	1	1		3	3	.....	2	2	0 or 1
7	1	1		1	1	.....	2	3	0 or 1
8	1	1		2	2	.....	3	1	0 or 1
9	1	1		3	3	.....	1	2	0 or 1
10	1	2		1	1	.....	3	2	0 or 1
11	1	2		2	2	.....	1	3	0 or 1
...	...	...	...	...	...	.....	...	...	...
32	2	2		2	1	.....	2	2	0 or 1
33	2	2		3	2	.....	3	3	0 or 1
34	2	2		1	3	.....	3	1	0 or 1
35	2	2		2	1	.....	1	2	0 or 1
36	2	2		3	2	.....	2	3	0 or 1

is satisfactory and 1 means the software has some problems or there are some bugs.

11.2.1 Study of Two-factor Combinations

Since the L_{36} array contains two-level and three-level factors, the following three types of two-factor combinations have to be studied:

1. Two-level factors
2. Three-level factors
3. Two-level and three-level factors

For factors A and B, the total number of a particular two-factor combination in an OA can be obtained by the following equation:

$$n_{ij} = (n_i * n_j)/N \longrightarrow \quad (11.1)$$

where n_{ij} = the number of combinations of ith level of A and jth level of B
n_i = the number of ith levels of A in a column that is assigned to A
n_j = the number of jth levels of B in a column that is assigned to B
N = total number of experimental runs in the array

11.2.2 Construction of Combination Tables

Let us explain procedure to construct combination tables by considering the two-level factors A and B. These factors are assigned to the columns 1 and 2 of the OA. For these factors, the possible combinations are A1B1, A1B2, A2B1, and A2B2, where A1 and A2 correspond to the first and second level of factor A and B1 and B2 are the first and second levels of the factor B. The number of these combinations can be obtained by equation (11.1). For this example, the number of combinations of A1B1 is equal to (18 × 18)/36, which is 9. Similarly, the number of other combinations is also equal to 9.

For obtaining combination effects, we have to check how many times the system failed in a given combination. If the system fails at all times, then there is something wrong with that combination and the designer needs to fix this combination. In this example, for a particular combination if the system fails all the time (9 times), then the designer has to look into this combination and take appropriate corrective measures. Since the responses for the combinations of the OAs are 0s or 1s, the number of 1s will determine the combinations to be fixed. The number of ones can be obtained by constructing Tables 11.2 and 11.3. In L_{36} array, the number of such tables is $^{11}C_2$, which is 55. For the two factors A and B, such a table is given as Table 11.2.

In the similar way, combination tables for three-level factors and combination tables for two-level and three-level factors can be constructed. Examples of three-level factors and two-level and three-level factors are shown in Tables 11.3 and 11.4, respectively.

Table 11.2 Interaction Table for Two-level Factors

Factor A / Factor B	*B1*	*B2*
A1	# of 1s	# of 1s
A2	# of 1s	# of 1s

Table 11.3 Interaction Table for Three-level Factors

Factor L / Factor M	*M1*	*M2*	*M3*
L1	# of 1s	# of 1s	# of 1s
L2	# of 1s	# of 1s	# of 1s
L3	# of 1s	# of 1s	# of 1s

Table 11.4 Interaction Table for Three-level Factors

Factor A/Factor W	*W1*	*W2*	*W3*
A1	# of 1s	# of 1s	# of 1s
A2	# of 1s	# of 1s	# of 1s

The total number of two-factor combinations for L_{36} array is = 55(two-level factors) + 66(three-level factors) + 132(two-level and three-level factors) = 253. Thus, using L_{36} array, we can study all the 1,606 combinations by conducting only thirty-six experiments and constructing the combination tables.

To summarize, the following steps outline the procedure for a system testing:

1. Identify active signals and their levels.
2. Select a suitable OA for testing and performing required experimental runs.
3. Construct the combination tables and identification of key combinations.

Even if the number of active signals is large, orthogonal arrays of higher size can be constructed to accommodate the signals.

11.3 MTS SOFTWARE TESTING (CASE STUDY 1)

Since software design is a good example of system design, we will show how this testing methodology is useful in testing software. This particular software is intended to perform a multivariate data analysis called Mahalanobis–Taguchi Strategy (MTS). MTS is a pattern analysis tool, which is useful to recognize and evaluate various patterns in multidimensional cases. Examples of multivariate systems are medical diagnosis systems, face/voice recognition systems, and inspection systems.

Since in this technique, Mahalanobis distance and Taguchi robust design principles are integrated, it is known as MTS.

Basically, there four stages in MTS. (Taguchi and Jugulum, 2002). They are:

Stage I: Construct a measurement scale

- Select a reference group with suitable variables and observations that are as uniform as possible. The reference group is also known as the Mahalanobis space (MS).
- Use the reference group as a base or reference point of the scale.

Stage II: Validate the measurement scale

- Identify the conditions outside the reference group.
- Compute the Mahalanobis distances of these conditions and check if they match the decision maker's judgment.
- Calculate signal-to-noise ratios (S/N ratios) to determine accuracy of the scale.

Stage III: Identify the useful variables (developing stage)

- Find out the useful set of variables using various combinations of variables with help of orthogonal arrays and S/N ratios.

Stage IV: Conduct future diagnosis with useful variables

- Monitor the conditions using the scale, which is developed with the help of the useful set of variables. Based on the values of Mahalanobis distances, appropriate corrective actions can be taken.

In MTS, the Mahalanobis distance (MD) can be calculated by using the following equation.

$$MD = D^2 = (1/k)Z_i C^{-1} Z_i^T \tag{11.2}$$

where Z_i = standardized vector obtained by standardized values of X_i ($i = 1, \ldots, k$)

$Z_i = (X_i - m_i)/s_i$;

X_i = value of ith characteristic

m_i = mean of ith characteristic

s_i = s.d of ith characteristic

k = number of characteristics/variables

T = transpose of the vector

C = correlation matrix

A detailed description of MTS method, along with case studies, is given in Chapter 12. Readers are encouraged to review this chapter for more details of the MTS method.

The software has been tested by using orthogonal arrays. As mentioned earlier, the most important aspect of this type of testing is selection of suitable usage conditions. Since this software is intended to perform a particular type of multivariate analysis, five usage conditions (factors) were selected:

1. Operating system (OS)
2. Number of variables
3. Sample size
4. Correlation structure
5. Data type

For these conditions, suitable levels were selected, as shown in Table 11.5.

Table 11.5 Factors and Levels for Testing

	Factors		*Level 1*	*Level 2*	*Level 3*
A	OS		WXP	WNT	
B	# of variables		17	60	98
C	Sample size		50	200	500
D	Correlations		Weak	Mild	Strong
E	Data type		Qualitative (2 levels)	Qualitative (10 levels)	Continuous

Since we have four factors at three levels and one factor at two levels, an L_{18} $(2^1 \times 3^7)$ array with 18 experimental combinations was selected for testing. The layout for this experimentation, along with experimental results, is shown in Table 11.6.

In Table 11.6 for columns where the factors A, B, C, D and E are assigned, 1, 2, and 3 are used to denote level 1, level 2 and level 3, respectively. In the column for test results, 0 indicates satisfactory performance and 1 indicates failure. The data were analyzed by evaluating all two-factor effects. The details of the analysis are shown in Table 11.7.

From this table, we can see that B2D3, B2E1, B3 C1, C1D2, C3D3, and D3E1 have 100 percent failures. These failures are related to higher number of variables, lower sample sizes, and strong correlations (multicollinearity problems). For example, B2D3 is a case with higher number of variables and strong correlations, and D3E1 is case of strong correlations with qualitative variables at two levels. With these problems the Mahalanobis distances cannot be computed precisely because the correlation matrix tends to become singular, causing problems for inverting the matrix. After identifying these problems, necessary actions were taken to change the algorithm for computing the Mahalanobis distances. This resulted in significant reduction in the number of bugs. Now the program works successfully in all eighteen combinations of OA.

Table 11.6 Experimental Layout and Results

	L18 Design					
	OS	*# of Variables*	*Sample Size*	*Correlations*	*Qualitative Data*	
Run #	*A*	*B*	*C*	*D*	*E*	*Test Result*
1	1	1	1	1	1	0
2	1	1	2	2	2	0
3	1	1	3	3	3	1
4	1	2	1	1	2	0
5	1	2	2	2	3	0
6	1	2	3	3	1	1
7	1	3	1	2	1	1
8	1	3	2	3	2	0
9	1	3	3	1	3	0
10	2	1	1	3	3	0
11	2	1	2	1	1	0
12	2	1	3	2	2	0
13	2	2	1	2	3	1
14	2	2	2	3	1	1
15	2	2	3	1	2	0
16	2	3	1	3	2	1
17	2	3	2	1	3	0
18	2	3	3	2	1	0

11.4 CASE STUDY 2

This study was conducted by a software company and demonstrates how software performance can be efficiently evaluated using this approach. As mentioned before, this software system is a good example to describe the method of system testing.

The software performance was required to be analyzed with twenty-three signals. These signals were numbered as A,B,C.., U,V,W. For these factors, suitable levels were selected. Table 11.8 shows some of the signal factors with chosen levels.

Table 11.7 Data Analysis (Two-factor Combinations in Percent Failures)

	B1	*B2*	*B3*		*C1*	*C2*	*C3*		*D1*	*D2*	*D3*		*E1*	*E2*	*E3*
A1	33.33	33.33	33.33		33.33	0.00	66.67		0.00	33.33	66.67		66.67	0.00	33.33
A2	0.00	66.67	33.33		66.67	33.33	0.00		0.00	33.33	66.67		33.33	33.33	33.33
				B1	0.00	0.00	50.00		0.00	0.00	50.00		0.00	0.00	50.00
				B2	50.00	50.00	50.00		0.00	50.00	100.00		100.00	0.00	50.00
				B3	100.00	0.00	0.00		0.00	50.00	50.00		50.00	50.00	0.00
								C1	0.00	100.00	50.00		50.00	50.00	50.00
								C2	0.00	0.00	50.00		50.00	0.00	0.00
								C3	0.00	0.00	100.00		50.00	0.00	50.00
												D1	0.00	0.00	0.00
												D2	50.00	0.00	50.00
												D3	100.00	50.00	50.00

Table 11.8 Signal Factors and Number of Levels

Signal	*A*	*L*	*M*	*N*	*O*	*P*	*B*	.....	*W*	*K*
Number of levels	2	3	3	3	3	3	2	.....	3	2

Table 11.9 Results of the Different Combinations of L_{36} Array

Expt. no	*1*	*2*	*3*	*4*	*5*	*6*	*7*	*8*	*9*	*10*	*11*	*12*	*13*	*14*	*15*	*16*	*17*	*18*
Response	1	0	0	0	1	1	0	1	0	0	1	1	0	0	0	0	0	0
Expt. no	19	20	21	22	23	24	25	26	27	28	29	30	31	32	33	34	35	36
Response	0	0	0	1	0	1	0	0	0	0	1	0	1	0	0	0	0	0

0: performance OK; 1: performance not OK.

The factors A,B,C,..,U,V,W were assigned to the different columns of L_{36} array as described before. The results of the thirty-six combinations are shown in Table 11.9.

11.4.1 Analysis of Results

With the help of the results of Table 11.10, the two-way combination tables were constructed. As mentioned before, for the signals in L_{36} array, the total number of two-way tables is 253. Out of all the two-factor combinations, only two combinations were considered important, as these had 100 percent errors. These combinations are K2 W1 and Q1 S1. These combinations are shown with arrows in Table 11.10. The combinations of K and W and Q and S are separately shown in Tables 11.11 and 11.12.

In Table 11.11, the different combinations of K and W are obtained from L_{36} array.

The combinations of Table 11.12 are obtained in a similar way.

11.4.2 Debugging the Software

After identifying the significant interactions, suitable corrective actions were taken. Then thirty-six more runs of the L_{36} array were conducted. In these runs, all the responses were 0, indicating that there were no bugs in the software.

Table 11.10 Main Table Showing Two-way Interactions

	B 1	B 2	..	C 1	C 2	..	D 1	D 2	..	S 1	S 2	S 3	..	W 1	W 2	W 3
A 1	4	2		5	1		3	3						4	1	1
A 2	2	2		1	3		2	2						3	1	0
B 1				3	3		3	3						4	2	0
B 2				3	1		1	3						3	0	1
C 1							3	3						4	1	1
C 2							3	1						3	1	0
..																
..																
K 1														1	0	0
K 2														6	2	1
..																
..																
Q 1										4	0	0				
Q 2										1	1	0				
Q 3										1	1	2				
..																
S 1																
S 2																
S 3																
..																
V 1														3	0	1
V 2														3	1	0
V 3														1	1	0

Table 11.11 Combinations of K and W

	W1	*W2*	*W3*	*Total*
K1	1	0	0	1
K2	6	2	1	9
Total	7	2	1	10

Table 11.12 Combinations of Q and S

	S1	*S2*	*S3*	*Total*
Q1	4	0	0	4
Q2	1	1	0	2
Q3	1	1	2	4
Total	6	2	2	10

11.5 CONCLUSIONS

- Principles of robust engineering can be applied for efficient system testing to eliminate most of the bugs.
- Orthogonal arrays are useful for testing to reduce test cases.
- OAs are used to test the performance in various combinations of usage conditions by studying all two-factor combinations. OAs enable us to foresee functioning of a product in various usage conditions, thus ensuring product reliability.
- OA method helps reduce most, but not all, of the bugs
- OA-based testing is simple and cost effective.

Chapter 12

Development of Multivariate Measurement System Using the Mahalanobis–Taguchi Strategy

In Six Sigma deployment, one of the most challenging aspects is to develop metrics to measure success of the system (it could be process, product, or service related). Quite often, we come across situations where we have to make decisions based on more than one variable or characteristic of the system. These systems are called *multivariate systems*. The examples of multivariate systems include medical diagnosis systems, manufacturing inspection system, face or pattern recognitions systems, fire alarm sensor systems, and so on. In this chapter, we describe Mahalanobis–Taguchi Strategy (MTS) and its applicability to develop multivariate measurement system. The Mahalanobis distance is used to measure the distances in a multivariate system, and Taguchi's principles are used to measure accuracy of the system and to identify important variables that are sufficient for the measurement system. This methodology is becoming increasingly popular, as there are more than 500 applications around the globe.

12.1 WHAT IS MAHALANOBIS–TAGUCHI STRATEGY?

Mahalanobis–Taguchi Strategy (MTS) is a pattern analysis technique, used to make accurate predictions through a multivariate measurement scale. Patterns are difficult to represent in quantitative terms, and they are extremely sensitive to correlations between the variables. The Mahalanobis distance (MD), which was introduced by a well-known Indian statistician P. C. Mahalanobis, measures distances of points in multidimensional spaces. The Mahalanobis distance has been extensively used in areas such as spectrographic applications and agricultural applications. This distance is proved to be superior to other multidimensional distances such as Euclidean distance because it takes correlations between the variables into account. Because of this, we use Mahalanobis distance (actually, a modified form of the original distance) to represent differences between individual patterns in quantitative terms. The Mahalanobis distance can be calculated as follows:

$$MD = D^2 = (1/k)Z_i C^{-1} {Z_i}^T \tag{12.1}$$

where Z_i = standardized vector obtained by standardized values of X_i ($i = 1 \ldots k$)

$Z_i = (X_i - m_i)/s_i$;

X_i = value of ith characteristic

m_i = mean of ith characteristic

s_i = s.d of ith characteristic

k = number of characteristics/variables

T = transpose of the vector

C = Correlation matrix.

Since computation of MD involves matrix theory, a brief discussion on topics related to matrix theory is provided in Appendix E.

For any scale, one must have a reference point from which the measurements can be made. Although it is easier to obtain a reference point for the scale with a single characteristic, it is not possible to obtain a single reference point when we are dealing with multiple characteristics. Therefore, in MTS the reference point corresponding to multiple variables is obtained with the help of a group of observations that are as uniform as possible and yet distinguishable through Mahalanobis distance. These observations are modified in such a way that their center is located at the origin (zero point) and the corresponding Mahalanobis distances are scaled so as to make average distance of this group unity. The zero point and unit distance thus obtained are used as reference point of the scale, and the distances are measured from this point. This set of observations is often referred to as *Mahalanobis space* (MS), *unit group*, or *normal group*. Selection of this group is entirely at the discretion of the decision maker conducting the diagnosis. In manufacturing-related applications, this group might correspond to parts having no defects; in medical diagnosis applications this group might consist of a set of people without any health problems; and in stock market predictions, this group could correspond to companies having average steady growth in a three-year period. The observations in this group are similar and not the same. Judicious selection of this group is extremely important for accurate diagnosis or predictions.

After developing the scale, the next step is the validation of the scale, which is done by the help of observations that are outside the Mahalanobis space. In this stage, we are, in a way, validating the Mahalanobis space (i.e., if it provides good base or reference point for future predictions/measurements); hence, the accuracy of the scale. This is important, because no method is considered good if it does not perform the intended function with observations that are not considered while developing the method. For the observations outside the Mahalanobis space, the distances are measured from the center of the normal group based on means, standard deviations, and correlation structure of this group. If the scale is good, the distances of these observations must match with the decision maker's judgment. In other words, if an observation does not belong to a normal group, then, it should have larger distance. Here we return to a measure called signal-to-noise ratio for assessing the accuracy of the scale. S/N ratio captures the correlation between the true or observed information (i.e., input signals) and the output of a system in the presence of uncontrollable variation (i.e., noise). In MTS, S/N ratio is defined as the measure of accuracy of predictions. A typical multidimensional predictive system that is used in MTS can be described using Figure 12.1.

As already mentioned, the output or prediction accuracy should have a good correlation with the input signal, and S/N ratios measure this correlation. The predictions are made based on the information on the variables defining the system, and they should be "accurate" even in the presence of noise factors such as different places of measurement, operating conditions, and so on. For example, in rainfall prediction, the input would be actual rainfall, and the output is the Mahalanobis distance

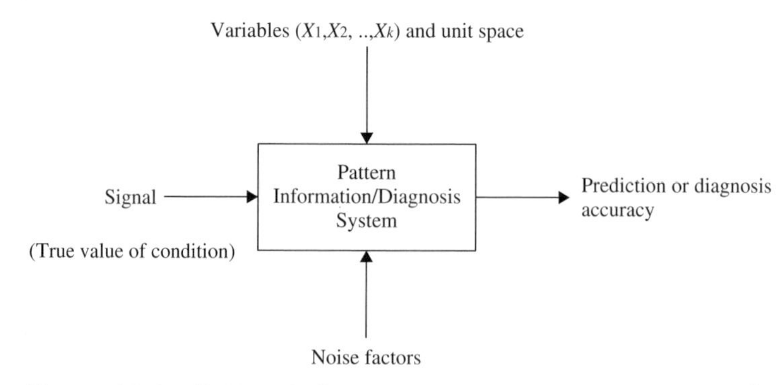

Figure 12.1 Pattern Information or Diagnosis System Used in MTS.

calculated based on the variables affecting the rainfall. In this case, S/N ratio measures correlation between actual rainfall and the Mahalanobis distance.

If the accuracy of predictions is satisfactory, then we will identify a useful subset of variables that is sufficient for the measurement scale while maintaining good prediction accuracy. Our experience shows that in many cases, the accuracy with useful variables is better than that with the original set of variables. However, in some cases the accuracy with useful variables might be less – which might be still desirable, as it helps reduce cost of inspection or measurement. In multidimensional systems, the total number of combinations to be examined would be of the order of several thousands, and hence it is not possible to examine all combinations. Here, we propose use of orthogonal arrays (OAs) to reduce the number of combinations to be tested. OAs are developed to estimate the effects of the variables by minimizing the number of combinations to be examined. They have been in use for quite a long time in the field of experimental design. In MTS, the variables are assigned to different columns of an OA. Based on S/N ratios obtained from different variable-combinations, important variables are identified. The future diagnosis is carried out only with these important variables.

12.2 STAGES IN MTS

Based on earlier discussion, the basic stages in MTS can be summarized as follows:

Stage I: Construct a measurement scale

- Select a Mahalanobis space with suitable variables and observations that are as uniform as possible.

- Use the Mahalanobis space as a base or reference point of the scale.

Stage II: Validate the measurement scale

- Identify the conditions outside the Mahalanobis space.
- Compute the Mahalanobis distances of these conditions and check if they match with the decision maker's judgment.
- Calculate S/N ratios to determine accuracy of the scale.

Stage III: Identify the useful variables (developing stage)

- Find out the useful set of variables using OAs and S/N ratios.

Stage IV: Conduct future diagnosis with useful variables

- Monitor the conditions using the scale, which is developed with the help of the useful set of variables.
- Based on the values of Mahalanobis distances, take appropriate corrective actions.

Figure 12.2 shows different steps in MTS.

From these discussion, it is clear that orthogonal arrays are prominent in the third stage of MTS analysis. Each experimental run in the orthogonal array design matrix uses a subset of variables; the resulting S/N ratios of these subsets are calculated using the distances outside the reference group, and S/N ratios are also used to determine the best variables.

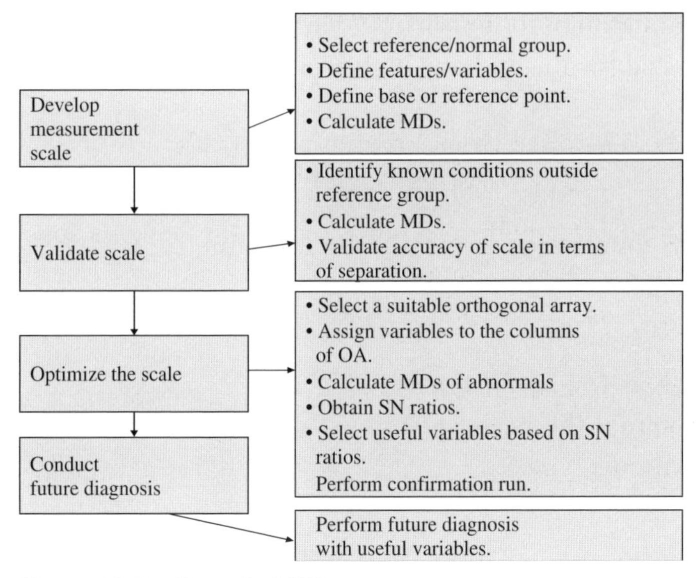

Figure 12.2 Steps in MTS.

12.3 SIGNAL-TO-NOISE RATIO – A MEASURE OF PREDICTION ACCURACY

In the context of MTS, signal-to-noise (S/N) ratio is used as a measure of prediction accuracy. As mentioned before, the accuracy of prediction with useful variables should be at least equal to that with all variables. Using S/N ratios will ensure a high level of prediction with useful variables. S/N ratios are computed for all combinations of OA based on MDs outside reference group. With these MDs, S/N ratios are obtained. Using S/N ratios as response, average effects of variables are computed at level 1 (presence) and level 2 (absence). Based on these effects, usefulness of variables can be determined. As we know, S/N ratio captures the magnitude of real effects (i.e., signals) after making some adjustment for uncontrollable variation (i.e., noise). Therefore, it is desirable to have high S/N ratios.

12.3.1 Types of S/N Ratios in MTS

In MTS application, typically the following types of S/N ratios are used:

- Larger-the-better type
- Nominal-the-best type
- Dynamic type

When the true levels of abnormals are not known, larger-the-better type S/N ratios are used if all the observations outside reference group are abnormals. This is because the MDs for abnormals should be higher. If the observations outside reference group are a mixture of normals and abnormals, then nominal-the-best type S/N ratios are used. When the levels of abnormals are known, dynamic S/N ratios are used.

Larger-the-better Type

The procedure for calculating S/N ratios corresponding to a run of an OA is as follows: Let there be t abnormal conditions. Let $D_1^2, D_2^2, \ldots, D_t^2$ be MDs corresponding to the abnormal situations. The S/N ratio (for larger-the-better criterion) corresponding to qth run of OA is given by:

$$\text{S/N ratio} = \eta_q = -10\text{Log}_{10}\left[(1/t)\sum_{i=1}^{t}(1/D_i^{\,2})\right] \tag{12.2}$$

Nominal-the-best Type

The procedure for calculating S/N ratios corresponding to a run of an OA is as follows: Let there be t abnormal conditions. Let $D_1^2, D_2^2, \ldots, D_t^2$ be MDs corresponding to the abnormal situations. The S/N ratio (nominal-the-better type) corresponding to qth run of OA is calculated as follows:

$$T = \text{Sum of all } D_i\text{s} = \sum_{i=1}^{t} D_i$$

$$S_m = \text{Sum of squares due to mean} = T^2/t$$

$$V_e = \text{Mean square error} = \text{variance} = \sum_{i=1}^{t} \frac{(D_i - \overline{D})^2}{(t-1)}$$

Where $\overline{D}$, is average of D_i

$$\text{S/N ratio} = \eta_{\text{q}} = 10Log_{10}\left[\frac{1/n(S_m - V_e)}{V_e}\right] \tag{12.3}$$

Dynamic Type

Examples of this type are weather forecasting systems and the case of rainfall prediction. The procedure for calculating S/N ratios corresponding to a run of an OA is as follows: Let there be t abnormal conditions. Let $D_1^2, D_2^2, \ldots, D_t^2$ be MDs corresponding to the abnormal conditions. Let $M_1, M_2, \ldots, M_t$ be true levels of severity (rainfall values in a rainfall prediction system example).

$$S_T = \text{Total sum of squares} = \sum_{i=1}^{t} {D_i}^2$$

$$r = \text{Sum of squares due to input signal} = \sum_{i=1}^{t} {M_i}^2$$

$$S_\beta = \text{Sum of squares due to slope} = (1/r)\left[\sum_{i=1}^{t} M_i D_i\right]^2$$

$$S_e = \text{Error sum of squares} = S_T - S_\beta,$$

$$V_e = \text{Error variance} = S_e/(t-1)$$

The S/N ratio corresponding qth run of OA is given by

$$\text{S/N ratio} = \eta_q = 10\log_{10}\{(1/r)[S_\beta - V_e]/V_e\} \tag{12.4}$$

12.4 MEDICAL CASE STUDY

To explain the applicability of MTS in medical applications, Dr. Kanetaka's (Taguchi and Jugulum, 2002) data on liver disease testing are used. The data contain observations of a healthy group, as well as of the abnormal conditions. The variables considered for the purpose of diagnosis are as shown in Table 12.1. The healthy group or the Mahalanobis Space (MS) is constructed based on observations on two hundred people who do not have any health problems. There are seventeen abnormal conditions. It is to be noted in this example that, by mere coincidence, the number of variables is equal to the number of abnormal conditions. This need not be so. For this system, different stages of MTS are applied as follows:

12.4.1 Stage 1: Development of Measurement Scale Using Mahalanobis Space

With the help of observations on two hundred healthy people, the MDs corresponding to all these people are computed. Mahalanobis space, in this case, is defined with the help of the MDs obtained for the healthy people. Table 12.2 gives the sample values of MDs of MS.

Table 12.1 Variables in Medical Diagnosis Data

S.No	*Variables*	*Notation*	*Notation for Analysis*
1	Age		X1
2	Sex		X2
3	Total protein in blood	TP	X3
4	Albumin in blood	Alb	X4
5	Cholinesterase	ChE	X5
6	Glutamate O transaminase	GOT	X6
7	Glutamate P transaminase	GPT	X7
8	Lactate dehydrogenase	LHD	X8
9	Alkanline phosphatase	Alp	X9
10	r-Glutamyl transpeptidase	r-GPT	X10
11	Leucine aminopeptidase	LAP	X11
12	Total cholesterol	TCh	X12
13	Triglyceride	TG	X13
14	Phospholopid	PL	X14
15	Creatinine	Cr	X15
16	Blood urea nitrogen	BUN	X16
17	Uric acid	UA	X17

12.4.2 Stage 2: Validation of the Measurement Scale

In this stage, the accuracy of the scale is justified by measuring the MDs of the known abnormal conditions. In this case, as mentioned before, there are seventeen known abnormal conditions. The MDs corresponding to these conditions are estimated. Table 12.3 gives the sample values of the abnormal conditions.

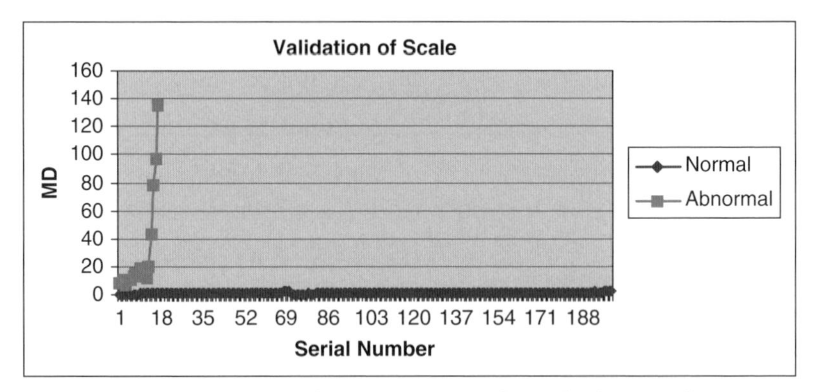

Figure 12.3 Separation between Normals and Abnormals (Validation of the Scale).

Since the MDs of abnormals have higher distances, the measurement scale is considered to be good.

Figure 12.3 clearly shows the separation between normals and abnormals, indicating ability of the scale to separate between normals and abnormals.

12.4.3 Stage 3: Identification of Useful Variables (Development Stage)

In this phase, a useful set of variables is identified using OAs and S/N ratios. Since there are seventeen variables and we required a two-level OA, $L_{32}(2^{31})$ array is selected. $L_{32}(2^{31})$ array is a two-level OA with thirty-two treatment combinations (runs) and thirty-one columns. The seventeen variables are allocated to the first seventeen columns of this array. The MDs corresponding to the abnormal conditions are computed for all thirty-two treatment combinations. For all combinations, MS is developed separately, with the number of variables present in the respective combinations.

Table 12.2 MDs corresponding to the Mahalanobis Space (Sample Values)

S. No	*1*	*2*	*3*	*4*	*5*	...	*198*	*199*	*200*	*Average*
MD	0.37837	0.43137	0.403562	0.5	0.5154	.....	1.776869	1.75632	2.358133	0.9951

Table 12.3 MDs Corresponding to Abnormal Conditions (Sample Values)

S. No	*1*	*2*	*3*	*4*	*5*	...	*15*	*16*	*17*
MD	7.727406	8.416294	10.29148	7.205157	10.59075	.....	78.6386	97.26777	135.6978

Table 12.4 Average Responses for Dynamic S/N Ratios (in dB Units)

	X1	*X2*	*X3*	*X4*	*X5*	*X6*	*X7*	*X8*	*X9*
Level 1	−8.18459	−8.18718	−8.24874	−7.9488	−7.069	−8.31781	−7.97568	−8.82373	−8.18751
Level 2	−7.74462	−7.74203	−7.68046	−7.9804	−8.86	−7.6114	−7.95353	−7.10548	−7.7417
Gain	−0.43997	−0.44515	−0.56828	0.0316	1.7912	−0.70641	−0.02215	−1.71825	−0.44581

	X10	*X11*	*X12*	*X13*	*X14*	*X15*	*X16*	*X17*
Level 1	−6.35846	−8.101	−7.8208	−7.56239	−7.31454	−7.58998	−7.96212	−7.83219
Level 2	−9.57075	−7.82821	−8.1085	−8.36682	−8.61467	−8.33922	−7.94709	−8.09702
Gain	3.21229	−0.27279	0.28769	0.80443	1.30013	0.74924	−0.03503	0.26483

Table 12.5 Sample Results (MDs) of the Confirmation Run

S. No	*1*	*2*	*3*	*4*	...	*14*	*15*	*16*	*17*	...	*198*	*199*	*200*
Healthy Group	0.37399	0.40438	0.5975	0.7975	...	0.7251	0.658	0.5399	0.6207	...	2.50897	1.25817	3.697
Abnormals	13.9373	14.7263	17.342	10.804	...	85.566	74.18	104.43	123.03				

Computation of S/N ratios

For all thirty-two combinations, dynamic S/N ratios are computed. As mentioned before, dynamic S/N ratios are used when there are different abnormal conditions with known levels of severity. These known levels of abnormals will act as signals to the system. Here, true levels of abnormals were not known, and hence working averages of the known groups are used as input signals. From the seventeen abnormal conditions, it was found that the first ten conditions belong to mild level of severity and the remaining seven belong to the medium level of severity. Therefore, the averages of square roots of MDs in each group are considered as different levels (M_1 and M_2) of input signal (M). The values of these levels are $M_1 = 3.388514$ and $M_2 = 6.917997$. The output response is MD corresponding to the abnormals. The S/N ratios are computed by using procedure given in Section 12.3.1. The values of S/N ratios of all runs of L_{32} (2^{31}) array can be obtained from the authors. The average responses corresponding to the seventeen variables are shown in Table 12.4.

In table 12.4, Level 1: Variable is present; Level 2: Variable is not present. From Table 12.4, it is clear that the variables X_4, X_5, X_{10}, X_{12}, X_{13}, X_{14}, X_{15}, and X_{17} have positive gains. That means these variables have higher average responses when they are part of the system (level 1). Hence, these variables are considered to be useful for future diagnosis process. The results of the confirmation run with useful variables showed that the measurement scale (developed with useful variables) can detect the abnormals. This can also be seen from Figure 12.4.

Table 12.5 gives sample results of the confirmation run. Moreover, the average MD of abnormal conditions with these variables is higher than that with all the variables. This means the insignificant variables will reduce the accuracy of MTS scale. Since the

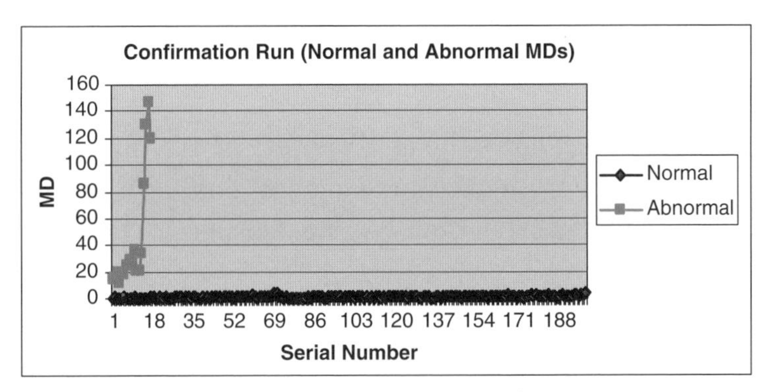

Figure 12.4 Normals and Abnormals (After Optimization).

gain in S/N ratio for X_4 is very small, a combination of the useful variables excluding X_4 is run to check the performance of the system. It is found that abnormals have lower MDs with this combination. Therefore, it has been decided to retain X_4 in the set of useful variables.

Gain in S/N Ratio

In case of dynamic analysis, the optimal combination is X_4-X_5-X_{10}-X_{12}-X_{13}-X_{14}-X_{15}-X_{17}. The S/N ratio for this combination is compared with the original combination. The details are given in Table 12.6. In this table, a gain of 1.9824 dB units indicates that the performance is better after optimization.

Similar analysis has been performed by using the larger-the-better type S/N ratios. The gain in S/N ratio, in this case, is found to be 1.45 dB units. This clearly shows that the dynamic type analysis gives better results.

Table 12.6 S/N Ratio Analysis (Dynamic Type)

S/N ratio-optimal system	– 4.26963 dB
S/N ratio-original system	– 6.25202 dB
Gain	1.9824 dB

12.5 CASE EXAMPLE 2: AUTO MARKETING CASE STUDY

This case example shows the applicability of MTS in marketing environment to help identify customers' buying patterns and characteristics that drive their buying decisions.

12.5.1 Introduction

This study is related to an auto marketing application where it is required to identify the customers' buying pattern of different car segments. The objective of this study is to recognize buying patterns of customers owning a particular model. Because of this objective, this is considered a pattern recognition application. The recognition of various patterns using MTS/MTGS analysis can be done as follows:

1. Construct the Mahalanobis space for a pattern under consideration (base pattern).
2. Consider other patterns as abnormals (conditions outside MS).
3. Select the useful variables by using orthogonal arrays and S/N ratios.
4. Use the useful variables for future diagnosis.
5. If we have prior knowledge about the abnormals, then recognize patterns by comparing them with the base pattern.
6. Otherwise repeat the steps 1 to 4 for all other patterns and test the new observation against all to decide which pattern it belongs to.

In this study, the patterns of the buyers were to be identified based on customer survey results. The variables considered for the survey are classified under three categories:

1. Personal views
2. Purchase reasons
3. Demographics

The customer survey data are obtained for these variables. After combining the variables in the three categories, fifty-five variables are considered. The number of car segments is five. The list of these variables can be obtained from the authors upon request.

In some cases, the customers were asked to rank them on a scale of 1 to 4, where 1 means strongly agree and 4 means strongly disagree. After arranging the fifty-five variables in a desired order, they are denoted as $X_1, X_2, .., X_{55}$ for the purpose of analysis. Since there are five segments and we did not have any prior knowledge about these patterns, MTS analysis is performed on all of the segments. For convenience, the five segments are denoted as $S_1, S_2, .., S_5$.

12.5.2 Construction of Mahalanobis Space

For all of the five segments, the Mahalanobis Space (MS) is constructed based on a huge data set. For example, MS for the S_1 is constructed by taking observations on fifty-five variables corresponding to that segment. With these Mahalanobis spaces, the corresponding MDs are calculated.

12.5.3 Validation of the Measurement Scale

The second stage in MTS methods is validation stage. The outside conditions for a given segment are chosen as conditions corresponding to the other segments. It is found that the abnormals, in all cases, have higher MDs and, hence, the scale is validated.

12.5.4 Identification of Useful Variables

For this purpose, since we have fifty-five variables, the L_{64} (2^{63}) orthogonal array (OA) was chosen for analysis. The S/N ratios are computed based on the larger-the-better criterion, because prior information about abnormals was not available.

Table 12.7 provides the list of useful variables corresponding to all five segments under consideration. Since for each segment a suitable strategy is to be developed to increase the sales of the cars, it is decided to restrict the number of variables per segment to twenty. This is done because it is easier to work with twenty variables and make practicable recommendations. The selection of these variables was done on the basis of the magnitude of gain in S/N ratio. In Table 12.7, in case S_2, the number of useful variables is nineteen, because these are the only variables with positive gains.

With the useful set of variables, confirmation runs are conducted for all five segments. The results of the confirmation indicate that these variables are able to recognize the given patterns as effectively as in the case with all fifty-five variables. Figure 12.5 shows the recognition power of the useful variables in a given segment.

Table 12.8 gives improvement in the S/N ratios of the entire system for all five segments. The table also provides corresponding variability reduction range (VRR).

From Table 12.8, it is clear that the improvement in S/N ratios is not very significant. However, the reduced number of variables in the optimal system helps in reducing the complexity of the multidimensional systems and helps developing good strategies to improve sales.

Table 12.7 Useful Variables Corresponding to the Five Segments

S. no	S_1	*S/N Ratio Gain*	S_2	*S/N Ratio Gain*	S_3	*S/N Ratio Gain*	S_4	*S/N Ratio Gain*	S_5	*S/N Ratio Gain*
1	X_6	0.20	X_{54}	0.60	X_{52}	1.15	X_7	0.50	X_2	0.96
2	X_{23}	0.16	X_7	0.57	X_{54}	0.73	X_{47}	0.29	X_{47}	0.49
3	X_{15}	0.11	X_{52}	0.48	X_{27}	0.27	X_{31}	0.28	X_{55}	0.49
4	X_{18}	0.11	X_{26}	0.36	X_{24}	0.23	X_{41}	0.27	X_7	0.45
5	X_3	0.11	X_{41}	0.30	X_6	0.19	X_{27}	0.25	X_{40}	0.35
6	X_{52}	0.09	X_{47}	0.28	X_{44}	0.18	X_2	0.22	X_{10}	0.33
7	X_{35}	0.09	X_{25}	0.19	X_3	0.15	X_3	0.22	X_3	0.24
8	X_{19}	0.09	X_{27}	0.16	X_{47}	0.15	X_{21}	0.22	X_{25}	0.20
9	X_{40}	0.09	X_{13}	0.14	X_{25}	0.15	X_{24}	0.16	X_{24}	0.17
10	X_{27}	0.09	X_{18}	0.11	X_4	0.13	X_{22}	0.13	X_{21}	0.16
11	X_{14}	0.08	X_8	0.09	X_2	0.13	X_{15}	0.13	X_6	0.15
12	X_{47}	0.08	X_3	0.08	X_{40}	0.12	X_{54}	0.12	X_{26}	0.15
13	X_{22}	0.07	X_{14}	0.08	X_{26}	0.11	X_8	0.12	X_{18}	0.15
14	X_{12}	0.07	X_{24}	0.08	X_{10}	0.10	X_{23}	0.11	X_{54}	0.14
15	X_5	0.06	X_{40}	0.07	X_{14}	0.10	X_{35}	0.11	X_{29}	0.13
16	X_4	0.06	X_{31}	0.05	X_{42}	0.07	X_{52}	0.10	X_{53}	0.09
17	X_{10}	0.06	X_{29}	0.05	X_{55}	0.06	X_{48}	0.08	X_{49}	0.06
18	X_1	0.06	X_{44}	0.02	X_{41}	0.05	X_{40}	0.07	X_{44}	0.06
19	X_{26}	0.05	X_{48}	0.00	X_{35}	0.05	X_{30}	0.06	X_{35}	0.05
20	X_{11}	0.05			X_{31}	0.01	X_{44}	0.04	X_{42}	0.03

Table 12.8 Gain in S/N Ratios and Variability Reduction Range (VRR)

Segment	*S/N Ratio – Before*	*S/N Ratio – After*	*Gain*	*VRR(%)*
S_1	2.9	2.98	0.08	0.92%
S_2	2.57	2.62	0.05	0.57%
S_3	1.12	1.27	0.15	1.72%
S_4	1.74	1.79	0.05	0.57%
S_5	1.68	1.79	0.11	1.26%

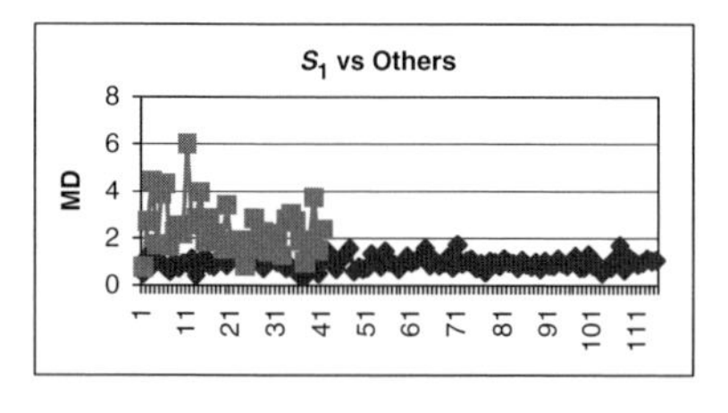

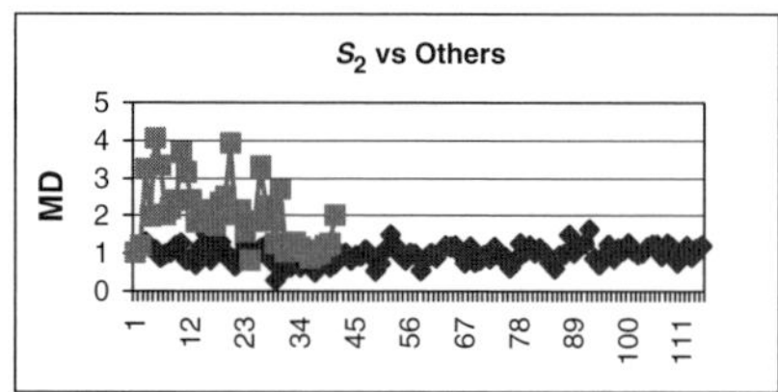

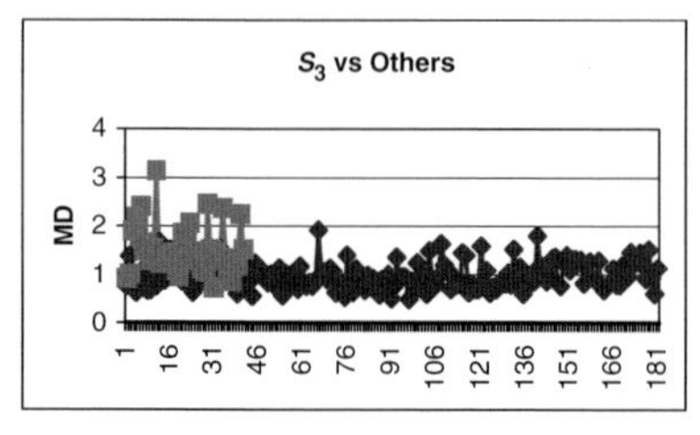

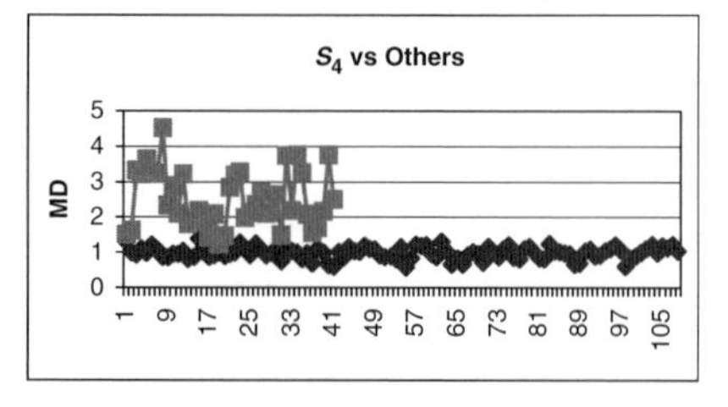

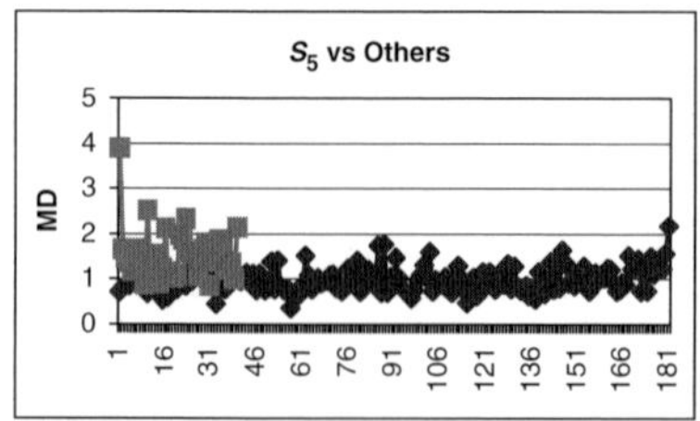

In these charts, plots with higher number of points correspond to respective segments and plots with less number of points correspond to others (from different segments).

Figure 12.5 Pattern Recognition with Useful Set of Variables.

12.6 CASE STUDY 3: IMPROVING CLIENT EXPERIENCE

This example is related to a financial institution. The summary of this study is described as follows.

12.6.1 Methodology

This project focused on improving the measurement of key indicators of client health. The client health was decided based 49 variables and classified into green (loyal), yellow (vulnerable), and red (at risk) groups of clients. The decisions were made based on client relationship manager's judgment without having any quantitative method, and hence the current method for measuring client relationship health was seen as subjective and not able to identify/differentiate specific types of risk from the primary client risk factors. The project primary goal is to develop a metric to quantitatively determine healthiness of a client based on the specified attributes.

The MTS methodology was applied to quantitatively determine client health. With all forty-nine variables, the MTS scale was constructed and validated. These are shown in Figure 12.6.

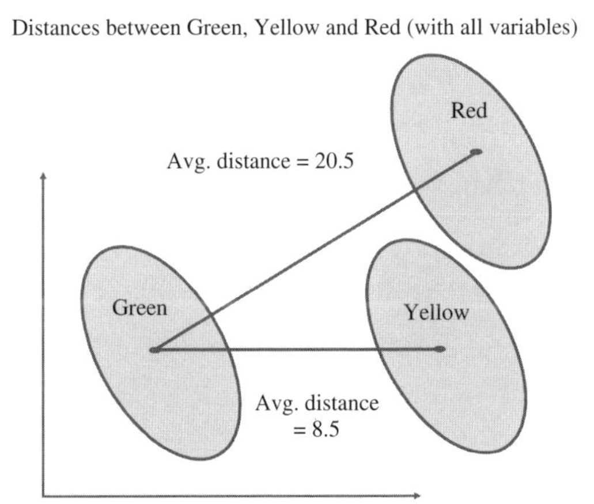

Figure 12.6 Validation of Scale.

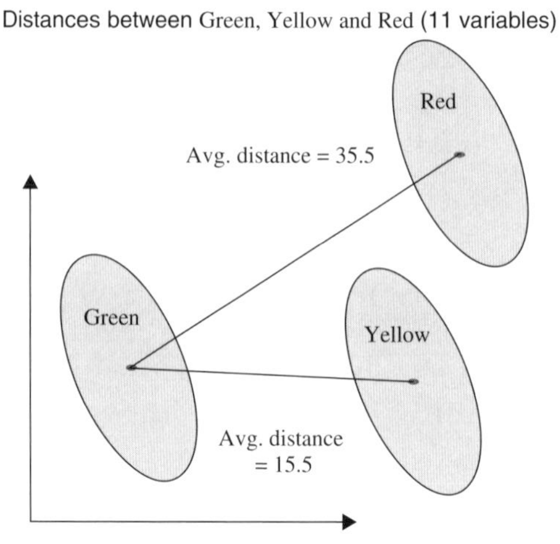

Figure 12.7 Separation between Clients (after Optimization).

After performing MTS optimization with L_{64} OA, and performing S/N ratio analysis and using other practical considerations, eleven variables were selected to be important. With these variables, the performance of the scale is as shown in Figure 12.7.

12.6.2 Improvements Made Based on Recommendations from MTS Analysis

A new measurement system has been developed by including the eleven attributes from the following:

- Survey attributes
- Volatility within client – merger, change to consultant, etc.
- Volatility within company – loss of one or more mandates, changes to staff, etc.

The company is convinced that the measurement system with these eleven attributes is much more meaningful and objective as compared to traditional subjective evaluations.

12.7 IMPROVEMENT OF THE UTILITY RATE OF NITROGEN WHILE BREWING SOY SAUCE

This case study describes the use of MTS method for improving utility of nitrogen while brewing soy sauce, or *tamari*. Tamari is a special type of thick soy sauce.

12.7.1 Introduction

In Japan, 50 percent of the total requirement of soy sauce and tamari is produced by five large-scale makers, and the remaining 50 percent is produced by 1,800 small and medium makers. In order to satisfy Japanese agricultural standards, soy sauce and tamari makers have consistently been producing a stable product. However, there is a great need to improve brewing technology in the case of small- and medium-scale manufacturers to compete with technology of large-scale manufacturers to maintain consistent quality and market share. In view of this, it has been decided to use MTS methodology to improve the brewing process. Since nitrogen content contributes significantly to the deliciousness of soy sauce, it is important to increase the utility rate of nitrogen during the brewing process. The utility rate of nitrogen is defined as the proportion of nitrogen, which has dissolved in *moromi*, unrefined soy sauce. The upper limit of the utility rate of nitrogen was fixed at 92 percent based on nitrogen that will not dissolve or be expelled as ammonia.

12.7.2 Process of Producing Soy Sauce or Tamari

There are four subprocesses in the production of soy sauce or tamari:

1. Material treatment
2. Preparation of koji-molds (special molds for brewage)
3. Aging
4. Inspection

The contributing factors in these sub processes are as follows:

- **Material treatment** – Type of materials, compounding, amount of water spraying, season for treatment, salt solution, condition of steaming and boiling, and so on.
- **Preparation of koji-molds** – Chamber of koji-molds, amount to be added, compounding, setting condition of air conditioner, temperature and so on.
- **Aging** – Temperature, agitation, condition of koji-molds.
- **Inspection** – Number of koji-molds bacteria, water content, power factor of molds-protein decomposing enzyme (PU30), pH, color, sediment, salt content, total nitrogen, alcohol, power factor of protein decomposing enzyme (PU15) in moromi (unrefined soy sauce), utility rate of dissolved nitrogen in moromi.

12.7.3 Selection of Factors for MTS Application

As mentioned before, the deliciousness of soy sauce significantly depends on the nitrogen content. Soy sauce contains 1.5 to 1.7 percent nitrogen and 2 to 3 percent alcohol, while tamari contains 2 to 3 percent nitrogen and 0.5 percent alcohol. Since manufacturing factors of soy sauce and tamari are common, the data corresponding to them are not treated separately. In other words, data corresponding to soy sauce and tamari are combined. During the process of producing soy sauce or tamari, the subprocesses koji-mold preparation and aging are considered to be critical. Hence, MTS method is separately applied to these subprocesses.

There are forty factors for koji-mold and eighty-two for aging. For these factors, the historical data were collected for a 3-year period.

12.7.4 MTS for Aging

The Mahalanobis space is generated based on upper limit on the utility rate of nitrogen. MS is generated based on 203 observations on 82 factors. Using MS, the measurement scale is developed with the help of MDs of normals. After this step, the scale is validated with known conditions. After performing dynamic S/N ratio analysis, the number of factors is reduced to 8 from 82. These factors are: f2, K1ge, k5, pu15, PH5end, iro3, roux, S, al. The factor pu15, which is a power factor of protein-decomposing enzyme in *moromi* resulted from koji-molds, is an important factor for improving the preparation process of koji-molds.

12.7.5 MTS for Koji-molding

The factor pu 15 in koji-molds is correlated with water content in koji-molds. As water content increases, the pu 15 will also be high. Since the measurement of water content is easier than measurement of pu 15, the MS was constructed based on the water content. The data for MS are obtained from a sample of 614 on 40 factors. After validating the measurement scale, dynamic S/N ratio analysis was conducted to identify a useful set of variables. As in aging, the number of useful factors are found to be eight. These factors are: rr, muro, m.wash, kisetu, mori.h, c.up.h, c.up.ond, and t2.10 h.

The useful factors obtained through MTS are sufficient to maintain the required utility rate of nitrogen and hence good quality of soy sauce or tamari. The method of MTS is included in the ISO 9001 quality system of the company that did this study.

12.8 APPLICATION OF MTS FOR MEASURING OIL IN WATER EMULSION

This study describes the use of MTS to predict the "healthiness" of oil in water emulsion.

12.8.1 Introduction

Sensitive materials, such as negative color film, consist of very thin sensitization layers. They make use of various other materials, and each of them provides specific function to satisfy the requirements of sensitive materials.

The sensitization layers use gelatin as a binder, and a water-soluble material can be directly added to sensitive materials as an aqueous solution. It is well known that an oil-soluble organic material, which is soluble in a solvent, can be added to sensitive materials with *drops of oil* (oil in water emulsion) by using surface active agents. Whether oil-soluble material provides its function in sensitive materials depends on the properties of drops of oil. The precipitation of materials inside drops of oil or rise ingrain size of drops of oil by own coalescence may result in the reduction of the performance of the materials. Therefore,

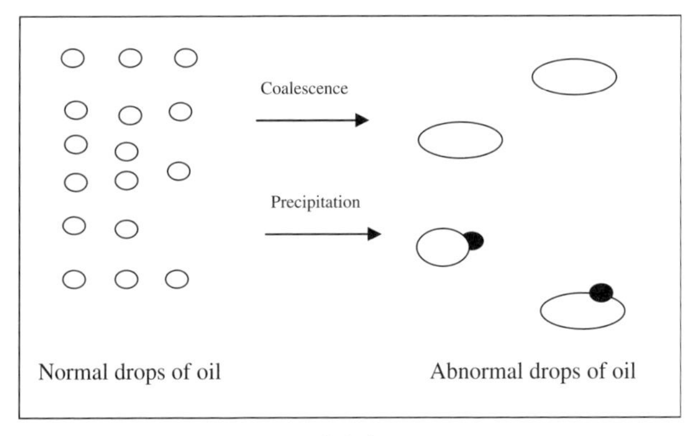

Figure 12.8 "Drops of Oil."

it is necessary to optimize type, quantity of many materials, solvent, and surface active agents to design drops of oil so that precipitation or coalescence does not occur.

To measure the conditions of drops of oil and the variable affecting the conditions of the same, MTS methodology is used. The useful variables should be able to predict the conditions of precipitation and coalescence. Figure 12.8 shows typical conditions of drops of oil.

The different conditions of drops of oil are referred to as different recipes.

12.8.2 Application of MTS

After defining the normal and abnormal conditions, the MTS method is systematically applied. The Mahalanobis space is constructed from the historical data corresponding to the recipes having no problems, such as precipitation and coalescence. The variables defining these recipes include various additives, solvents, and surface active agents. From MS, the MDs are computed for normals. For known abnormals, MDs are computed to validate the measurement scale. Here, the abnormal conditions are generated by carrying out experiments. It was found that the normal recipes have MDs of 5.0 or less and abnormals have very high MDs. It was found that for recipes with MDs of 5.0 or less, the probability of occurrence of coalescence is very low. This indicates that for these recipes, the need for conducting compulsive experiments greatly reduced and hence saves man hours and cost. This shows the power of MD to predict failures while minimizing total cost. In the next step, the useful variables are identified with help of S/N ratios to predict failures in a more efficient manner.

After developing MTS scale with useful variables, it is applied to actual recipe designs. The MDs corresponding to these recipes were able to predict the conditions accurately. This was ascertained by carrying out detailed experimentation.

Thus, MTS method is very helpful in predicting abnormal recipes so that appropriate corrective actions can be taken. This helps in using the right kind of recipe so that the performance of sensitive materials is not affected. This helps in reducing unnecessary experimentation and man hours.

12.9 PREDICTION OF FASTING PLASMA GLUCOSE (FPG) FROM REPETITIVE ANNUAL HEALTH CHECK-UP DATA

This case study describes the use of MTS technique for predicting fasting plasma glucose (FPG) to control diabetes mellitus.

12.9.1 Introduction

In a survey conducted by Japanese Ministry of Health and Welfare in 1997, it was reported that about 6.9 million Japanese have diabetes and about 13.7 million Japanese have blood glucose different from the standard. The blood glucose content can be used as a measure of diabetes mellitus content. Diabetes mellitus largely depends on day-to-day activities (like food habits and exercise habits). When detected at early stage, diabetes mellitus can be controlled by improving day-to-day activities. As diabetes mellitus depends on several variables, MTS method is used to predict the same through FPG. The prediction results are also compared with stepwise regression analysis.

12.9.2 Diabetes Mellitus

Diabetes mellitus is a group of diseases characterized by chronic hyperglycemia based on the degree of deficiency of insulin effect, decrease in insulation secretion, metabolic abnormalities, and chronic hyperglycemia. The Japanese Diabetes Society provided the following standard criterion to make a decision on these diseases.

Diabetes FPG $\geq$ 126mg/dl

Normal FPG $\geq$ 110mg/dl

Border those who are in between diabetes and normal

For diagnosing these diseases, one has to ascertain FPG values at least two times.

12.9.3 Application of MTS

The normal and abnormal conditions are defined in accordance with standards of the Diabetes Society. Various factors such as age, diabetes family history, food habits, exercise habits, corpulence degree (BMI), and blood pressure are considered for MTS analysis.

Initially, data on these variables were collected for one year. To improve the accuracy of prediction, the number of observations was increased by adding subsequent years of data. This

Table 12.9 A Comparison of MTS Based Prediction with Traditional Multivariate Analysis

	FPG Value
MTS-based prediction	134.3 ± 7.2 mg/dl
Prediction from traditional multivariate analysis	117.7 ± 6.2 mg/dl
Measured value	132.7 ± 9.0 mg/dl

process was stopped after collecting five years of data. From this data, a Mahalanobis space was constructed. After validating the measurement scale, S/N ratio analysis was conducted to find a useful set of variables. It was found that this useful variable set, thus obtained, is sufficient for future predictions of FPG (and hence diabetes mellitus) based on MDs. As shown in Table 12.9, the predictions based on MD are found to be better than those with stepwise regression analysis. In other words, MTS based predictions are much closer to the measured values than predictions based on traditional multivariate analysis.

Thus, the MTS method is very useful in predicting diabetes mellitus, which will enable patients to take appropriate preventive actions.

References

Allen, J. K., C. C. Seepersad, and F. Mistree. "*A Survey of Robust Design with* Applications to Multidisciplinary and Multiscale Systems." *Journal of Mechanical Design*, special issue on Risk-based and Robust Design. Ed. by S. Azarm and Z. Mourelatos, 128, 832–843, 2006.

Altshuller, G. S. *Creativity as an Exact Science: The Theory of the Solution of Inventive Problem Solving*. CRC Publisher, Boca Raton, FL, 1984.

Anderson, T. W. *An Introduction to Multivariate Statistical Analysis* 2nd ed. New York: John Wiley, 1984.

Blair, R. "Optical Fiber Insensitive to Temperature Variations." U.S. Pat. #4,432,606. 1984.

Box, G. E. P. "Statistics as a Catalyst to Learning by the Scientific Method Part II – a Discussion." *Journal of Quality Technology* 31(1) (1999): 16–29.

Brown, William C. *Matrices and Vector Spaces*. New York, NY. Marcel Dekker, Inc., 1991.

Brue, Greg. *The Design for Six Sigma*. New York: McGraw-Hill, 2003.

Chang, Victor, O. Chang, and M. Campo. "System for Measuring Flow." U.S. Pat. #5,483,840. 1996.

Christensen, C., S. Anthony, G. Berstell and D. Nitterhouse. 2007. Finding the Right Job for Your Product, *Sloan Management Review*, Spring 2007, Vol. 48, No. 3, pp. 38–47, MIT.

Clausing, Don. *Total Quality Development*. New York: ASME Press, New York, NY. 1994.

Clausing, D. P., and V. Fey. *Effective Innovation*. New York: ASME Press, 2004.

Clausing, D. P., and D. D. Frey. "Improving System Reliability by Failure-mode Avoidance Including Four Concept Design Strategies." *Systems Engineering* 8(3) (2005): 245–261.

Coombs, Clyde F. Jr. *Printed Circuits Handbook*. New York: McGraw-Hill, 1988.

Creveling, C. M., J. L. Slutsky, and D. Antis Jr. *Design for Six Sigma in Technology and Product Development*. Upper Saddle River, NJ: Prentice Hall, 2002.

Dasgupta, Somesh. 1993. "The Evolution of the D^2-statistic of Mahalanobis." *Sankhya* 55 (1993): 442–459.

De Moivre, A. 1730. *Miscellanea Analytica de Seriebus et Qudraturis*, London: J. Tonson and J. Watts.

Domb, E., and K. Tate. 1997. 40 Inventive Principles With Examples, *The TRIZ Journal*, July 1997.

Domb 1998. The 39 Features of Altshuller's Contradiction Matrix, *The TRIZ Journal*, November 1998.

Domb, E., J. Terninko, J. Miller and E. MacGran. 1999. "The Seventy-Six Standard Solutions: How They Relate to the 40 Principles of Inventive Problem Solving." *The TRIZ Journal*, May 1999.

Domb, E., et al. 2000a. "The Seventy-Six Standard Solutions, with Examples Section One." *The TRIZ Journal* (February 2000).

Domb, E., et al. 2000b. "The Seventy-Six Standard Solutions, with Examples Class Two." *The TRIZ Journal* (March 2000).

Domb, E., et al. 2000c. "The Seventy-Six Standard Solutions, with Examples Class Three." *The TRIZ Journal* (May 2000).

Domb, E., et al. 2000d. "The Seventy-Six Standard Solutions, with Examples Class Four." *The TRIZ Journal* (June 2000).

Domb, E., et al. 2000e. "The Seventy-Six Standard Solutions, with Examples Class Five." *The TRIZ Journal* (July 2000).

Domb, E. and K. Tate 2002. *Simplified TRIZ: New Problem-Solving Applications for Engineers & Manufacturing Professionals*, CRC Press.

Ealey, Lance A. *Quality by Design: Taguchi Methods and U.S. Industry*. Dearburn, MI, ASI Press, 1994.

Fey, V. R., and E. I. Rivin. *The Science of Innovation*. Southfield, MI: TRIZ Group, 1997.

Ford, R. B., and P. Barkan. "Beyond Parameter Design – A Methodology Addressing Product Robustness at the Concept Formation Stage." *Current Engineering and Design / Manufacturing Integration* 81 (1995): 1–7.

Ford, R. B. "Process for the Conceptual Design of Robust Mechanical Systems: Going Beyond Parameter Design to Achieve World-class Quality." Ph.D. diss., Stanford University, 1996.

Foster, R, and S. Kaplan. 2001. *Creative Destruction: Why Companies That Are Built to Last Underperform the Market–And How to Successfully Transform Them*, Currency.

Fowlkes, W. Y., and C. M. Creveling. *Engineering Methods for Robust Product Design*. New York: Addison-Wesley, 1995.

Freeman, C., and R. R. Moritz. "Gas Turbine Engine with Improved Compressor Casing for Permitting Higher Air Flow and Pressure Ratios before Surge." U.S. Pat. #4,086,022. 1978.

Frey, Daniel D., and Rajesh Jugulum. "Robustness through Invention." *Journal of Quality Engineering Society* 12(3) (2004): 116–122.

Gerlach, T. "A Simple Micropump Employing Dynamic Passive Valves Made in Silicon." *Micro System Technologies '94* (1994): 1025–1034.

Gerlach, T. "Fundamentals of Dynamic Flux Rectification as the Basis of Valve-Less Dynamic Micropumps." *Micro System Technologies '96* (1996): 445–450.

Guilford, J. P. 1950. Creativity, *American Psychologist*, 5, 444–454.

Hohn, Franz E. *Elementary Matrix Algebra*. New York: Macmillan, 1967.

IBM Global Business Services 2006. *Expanding the Innovation Horizon: The Global CEO Study 2006*, IBM Corporation.

Jiang, Lan, and Venkat Allada. "Robust Modular Product Family Design Using a Modified Taguchi Method." *Journal of Engineering Design* 16(5) (2005): 443–458.

Johnson, Richard A., and Dean W. Wichern. *Applied Multivariate Statistical Analysis*. Englewood Cliffs, NJ: Prentice Hall, 1992.

Jugulum, Rajesh, Shin Taguchi, and Kai Yang. "New Developments in Multivariate Diagnosis: A Comparison between Two Methods." *Journal of Japanese Quality Engineering Society* 7(5) (1999): 62–72.

Jugulum, Rajesh. "New Dimensions in Multivariate Diagnosis to Facilitate Decision Making Process." Ph.D. diss., Wayne State University, 2000.

Jugulum, Rajesh, and Daniel D. Frey. "Robustness through Invention." *Journal of Japanese Quality Engineering Society* (2002).

Jugulum, R., G. Taguchi, S. Taguchi, J. O. Wilkins, D. M. Hawkins, B. Abraham and A. M. Variyath. "Discussion of a Review and Analysis of the Mahalanobis–Taguchi System by Woodall, W. H. et al." *Technometrics* 45 (2003): 16–29.

Jugulum, R., and D. D. Frey. "Toward a Taxonomy of Concept Designs for Improved Robustness." *Journal of Engineering Design* 2 (2007): 139–156.

Kanetaka, Tatsuji. "Application of Mahalanobis Distance, Standardization, and Quality Control." *Japanese Standards Association* 41 (5 and 6) (1988).

Kim, I. Y., and A. Tezuka. "Optimization and numerical flow analysis of a valveless micropump." *JSME International Journal*, Series C: Mechanical Systems, Machine Elements and Manufacturing 46 (2003): 772–778.

Lacey, Damien, and Client Steele. "The Use of Dimensional Analysis to Augment Design of Experiments for Optimization and Robustification." *Journal of Engineering Design* 17(1) (2006) 55–73.

Lee, Taesik, and Rajesh Jugulum. Axiomatic Design: Advances and Applications, Solutions to Selected Problems. Unpublished work, 2003.

Mahalanobis, P. C. *On the Generalized Distance in Statistics*, Proceedings. National Institute of Science of India 2 (1936): 49–55.

Mann, D. *Hands on Systematic Innovation*. IFR Press, Clevedon, UK. 2002.

Matsumoto, S., A. Klein, and R. Maeda. "Development of Bi-directional Valve-less Micropump for Liquid." *IEEE MEMS '99* (1999): 141–146.

Matsumoto, S., R. Maeda, and A. Klein. "Characterization of a Valve-less Micropump Based on Liquid Viscosity." *Microscale Thermophysical Engineering* 3 (1999): 31–42.

Monplaisir, Leslie, Rajesh Jugulum, and Mahfoozulhaq Mian. Application of TRIZ and Taguchi Methods: Two Case Examples. *TRIZ Journal* (January 1999).

Monroe, R. C. "Axial Flow Fans and Blades Therefore." U.S. Pat #4,345,877. 1982.

Morrison, Donald F. *Multivariate Statistical Methods*. New York: McGraw-Hill, 1967.

Morrison, Donald F. Multivariate Statistical Methods. *Series in Probability and Statistics*, 3rd ed. New York: McGraw-Hill, 1990.

Nair, Vijayan N. Taguchi's Parameter Design: A Panel Discussion. *Technometrics* 34(2) (1993): 127–161.

Negrin, D. "Method and System for Detecting and Compensating for Rough Roads in an Anti-lock Brake System." U.S. Pat. #5,627,755. 1997.

Olsson, A., and G. Stemme. Micromachined Diffuser/Nozzle Elements for Valve-Less Pumps. *IEEE MEMS '96* (1996): 378–383.

Park, Sung H. *Robust Design and Analysis for Quality Engineering*. Chapman & Hall, 1996.

Phadke, Madhav S. *Quality Engineering Using Robust Design*. Englewood Cliffs, NJ: Prentice Hall, 1989.

Phadke, M. S., and Genichi Taguchi. "Selection of Quality Characteristics and S/N ratios for Robust Design." In conference Record, GLOBECOM 87 Meeting, IEEE Communication Society, Tokyo, Japan, 1987, 1002–1007.

Repple, W. O., and J. R. L. Fulton. "Coolant Pump for Automotive Use." U.S. Pat. #6,309,193. 2001.

Rojo, M. A. H., S. Santilan, and M. Sabelkin. "Redesign Improvement of Hydrocarbon Sampling Tool Using Robust Design." *Journal of Engineering Design* 16(1) (2005): 75–90.

Silverstein, D., M. Slocum and N. DeCarlo. 2005. *Insourcing Innovation: How to Achieve Competitive Excellence Using TRIZ*, Auerback.

Smith, Jonathan, and John Clarkson. "Design Concept Modeling to Improve Reliability." *Journal of Engineering Design* 16(5) (2005): 473–492.

Stemme, E., and G. Stemme. "A Valveless Diffuser/Nozzle-Based Fluid Pump." *Sensors and Actuators A*. 39 (1993): 159–167.

Suh., N. P. "Axiomatic Design of Mechanical Systems." *ASME Journal of Mechanical Design* 117 (1995): 2–10.

Suh., N. P. *Axiomatic Design: Advances and Applications*. New York: Oxford University Press, 2001.

Suh., N. P. *Complexity: Theory and Applications*. New York: Oxford University Press, 2005.

Sullivan, J. J., and J. D. Crofts. "Piezoelectric Actuator and Valve Assembly with Thermal Expansion Compensation." U.S. Pat. #6,313,568. 2001.

Taguchi, Genichi, and Jikken Kiekakuho. *Design of Experiments*, vols. I and II. Tokyo: Maruzen Co., 1976–77.

Taguchi, Genichi. *Introduction to Quality Engineering*. Tokyo: Asian Productivity Organization, 1986.

Taguchi, G. *System of Experimental Design*, vols. 1 and 2. White Plains, NY: ASI & Quality Resources, 1987.

Taguchi, G. "The Development of Quality Engineering." *The ASI Journal* 1(1) (1988): 5–29.

Taguchi, Genichi. *Taguchi on Robust Technology Development*. New York: ASME Press, 1993.

Taguchi, Genichi. "Diagnosis and Signal-to-Noise Ratio." *Quality Engineering Forum* 2(4 and 5) (1994).

Taguchi, Genichi. "Application of Mahalanobis Distance for Medical Treatment." *Quality Engineering Forum* 2(6) (1994).

Taguchi, Genichi, and Rajesh Jugulum. "Role of S/N Ratios in Multivariate Diagnosis." *Journal of Japanese quality engineering society* 7(6) (1999): 63–69.

Taguchi, Genichi, and Rajesh Jugulum. "Taguchi Methods for Software Testing." JUSE Software Quality Conference Proceedings, 2000.

Taguchi, Genichi, and Rajesh Jugulum. "New Trends in Multivariate Diagnosis." *Sankhya, Indian journal of statistics*, series B Part 2 (2000): 233–248.

Taguchi, Genichi, and Rajesh Jugulum. *The Mahalanobis-Taguchi Strategy: A Pattern Technology*. New York: John Wiley & Sons, 2002.

Taguchi, G., R. Jugulum, and S. Taguchi. *Computer-based Robust Engineering: Essentials for DFSS*. Milwaukee, WI: ASQ Quality Press, 2004.

Tentler, M. L., and G. L. Wheeler. "Viscosity-insensitive Variable-area Flowmeter." U.S. Pat. #5,024,105. 1991.

Tsai, You-Tern. *The preliminary investigation of system reliability and maintainability to develop availability sound designs*, 16(5), 459–471.

Ullman, D. G. *The Mechanical Design Process*. New York: McGraw-Hill, 1992.

Wallas, G. 1926. *The Art of Thought*, New York: Franklin Watts

Wall Street Journal. 2007. How Motorola Fell a Giant Step Behind: As It Milked Thin Phone, Rivals Sneaked Ahead on the Next Generation, *Wall Street Journal*, 98, April 2007

Womack, J., D. T. Jones and D. Ross. 1991. *The Machine That Changed the World: The Story of Lean Production*, Harper Perennial.

Womack, J. and Jones, D.T.. 2003. *Lean Thinking: Banish Waste and Create Wealth in Your Corporation*, Freepress

Wu, C. F. J., and M. Hamada. *Experiments: Planning, Analysis, and Parameter Design Optimization*. New York: Wiley & Sons, 2000.

Yang, Kai, Basem S. EI-Haik. *Design for Six Sigma: A Roadmap for Product Development*. New York: McGraw-Hill Professional, 2003.

Zook, C., J. Allen. 2001. *Profit from the Core: Growth Strategy in an Era of Turbulence*, Harvard Business School Press.

Appendixes

Appendix A

TRIZ Contradiction Matrix

	Worsening Feature → / Improving Feature ↓	Weight of moving object	Weight of stationary object	Length of moving object	Length of stationary object	Area of moving object	Area of stationary object	Volume of moving object	Volume of stationary object	Speed	Force (Intensity)	Stress or pressure	Shape	Stability of the object's composition	Strength	Duration of action of moving object	Duration of action of stationary object	Temperature	Illumination intensity	Use of energy by moving object	Use of energy by stationary object	Power	Loss of Energy	Loss of Substance	Loss of Information	Loss of Time	Quantity of substance	Reliability	Measurement accuracy	Manufacturing precision	Object-affected harmful factors	Object-generated harmful factors	Ease of manufacture	Ease of operation	Ease of repair	Adaptability or versatility	Device complexity	Difficulty of detecting and measuring	Extent of automation	Productivity
		1	2	3	4	5	6	7	8	9	10	11	12	13	14	15	16	17	18	19	20	21	22	23	24	25	26	27	28	29	30	31	32	33	34	35	36	37	38	39
1	Weight of moving object		-	15, 8, 29,34	-	29, 17,	-	29, 2, 40, 28	-	2, 8, 15, 38	8, 10, 18, 37	10, 36,	10, 14,	1, 35, 19, 39	28, 27,	5, 34, 31, 35	-	6, 29, 4, 38	19, 1, 32	35, 12,	-	12, 36,	6, 2, 34, 19	5, 35, 3, 31	10, 24, 35	10, 35,	3, 26, 18, 31	1, 3, 11, 27	28, 27,	28, 35,	22, 21,	22, 35,	27, 28, 1,	35, 3, 2, 24	2, 27, 28, 11	29, 5, 15, 8	26, 30,	28, 29,	26, 35 18, 19	35, 3, 24, 37
2	Weight of stationary object	-		-	10, 1, 29, 35	-	35, 30,	-	5, 35, 14, 2	-	8, 10, 19, 35	13, 29,	13, 10,	26, 39, 1,	28, 2, 10, 27	-	2, 27, 19, 6	28, 19,	19, 32, 35	-	18, 19,	15, 19,	18, 19,	5, 8, 13, 30	10, 15, 35	10, 20,	19, 6, 18, 26	10, 28, 8,	18, 26, 28	10, 1, 35, 17	2, 19, 22, 37	35, 22, 1,	28, 1, 9	6, 13, 1, 32	2, 27, 28, 11	19, 15, 29	1, 10, 26, 39	25, 28,	2, 26, 35	1, 28, 15, 35
3	Length of moving object	8, 15, 29, 34	-		-	15, 17, 4	-	7, 17, 4, 35	-	13, 4, 8	17, 10, 4	1, 8, 35	1, 8, 10, 29	1, 8, 15, 34	8, 35, 29, 34	19	-	10, 15, 19	32	8, 35, 24	-	1, 35	7, 2, 35, 39	4, 29, 23, 10	1, 24	15, 2, 29	29, 35	10, 14,	28, 32, 4	10, 28,	1, 15, 17, 24	17, 15	1, 29, 17	15, 29,	1, 28, 10	14, 15, 1,	1, 19, 26, 24	35, 1, 26, 24	17, 24,	14, 4, 28, 29
4	Length of stationary object		35, 28,	-		-	17, 7, 10, 40	-	35, 8, 2,14	-	28, 10	1, 14, 35	13, 14,	39, 37, 35	15, 14,	-	1, 10, 35	3, 35, 38, 18	3, 25	-		12, 8	6, 28	10, 28,	24, 26,	30, 29, 14		15, 29, 28	32, 28, 3	2, 32, 10	1, 18		15, 17, 27	2, 25	3	1, 35	1, 26	26		30, 14, 7,
5	Area of moving object	2, 17, 29, 4	-	14, 15,	-		-	7, 14, 17, 4		29, 30, 4,	19, 30,	10, 15,	5, 34, 29, 4	11, 2, 13, 39	3, 15, 40, 14	6, 3	-	2, 15, 16	15, 32,	19, 32	-	19, 10,	15, 17,	10, 35, 2,	30, 26	26, 4	29, 30, 6,	29, 9	26, 28,	2, 32	22, 33,	17, 2, 18, 39	13, 1, 26, 24	15, 17,	15, 13,	15, 30	14, 1, 13	2, 36, 26, 18	14, 30,	10, 26,
6	Area of stationary object	-	30, 2, 14, 18	-	26, 7, 9, 39	-		-		-	1, 18, 35, 36	10, 15,		2, 38	40	-	2, 10, 19, 30	35, 39, 38		-		17, 32	17, 7, 30	10, 14,	30, 16	10, 35, 4,	2, 18, 40, 4	32, 35,	26, 28,	2, 29, 18, 36	27, 2, 39, 35	22, 1, 40	40, 16	16, 4	16	15, 16	1, 18, 36	2, 35, 30, 18	23	10, 15,
7	Volume of moving object	2, 26, 29, 40	-	1, 7, 4, 35	-	1, 7, 4, 17	-		-	29, 4, 38, 34	15, 35,	6, 35, 36, 37	1, 15, 29, 4	28, 10, 1,	9, 14, 15, 7	6, 35, 4	-	34, 39,	2, 13, 10	35	-	35, 6, 13, 18	7, 15, 13, 16	36, 39,	2, 22	2, 6, 34, 10	29, 30, 7	14, 1, 40, 11	25, 26, 28	25, 28, 2,	22, 21,	17, 2, 40, 1	29, 1, 40	15, 13,	10	15, 29	26, 1	29, 26, 4	35, 34,	10, 6, 2, 34
8	Volume of stationary object	-	35, 10,	19, 14	35, 8, 2, 14	-		-		-	2, 18, 37	24, 35	7, 2, 35	34, 28,	9, 14, 17, 15	-	35, 34, 38	35, 6, 4		-		30, 6		10, 39,		35, 16, 32	35, 3	2, 35, 16		35, 10, 25	34, 39,	30, 18,	35		1		1, 31	2, 17, 26		35, 37,
9	Speed	2, 28, 13, 38	-	13, 14, 8	-	29, 30, 34	-	7, 29, 34	-		13, 28,	6, 18, 38, 40	35, 15,	28, 33, 1,	8, 3, 26, 14	3, 19, 35, 5	-	28, 30,	10, 13, 19	8, 15, 35, 38	-	19, 35,	14, 20,	10, 13,	13, 26		10, 19,	11, 35,	28, 32, 1,	10, 28,	1, 28, 35, 23	2, 24, 35, 21	35, 13, 8,	32, 28,	34, 2, 28, 27	15, 10, 26	10, 28, 4,	3, 34, 27, 16	10, 18	
10	Force (Intensity)	8, 1, 37, 18	18, 13, 1,	17, 19, 9,	28, 10	19, 10, 15	1, 18, 36, 37	15, 9, 12, 37	2, 36, 18, 37	13, 28,		18, 21, 11	10, 35,	35, 10, 21	35, 10,	19, 2		35, 10, 21	-	19, 17, 10	1, 16, 36, 37	19, 35,	14, 15	8, 35, 40, 5		10, 37, 36	14, 29,	3, 35, 13, 21	35, 10,	28, 29,	1, 35, 40, 18	13, 3, 36, 24	15, 37,	1, 28, 3, 25	15, 1, 11	15, 17,	26, 35,	36, 37,	2, 35	3, 28, 35, 37
11	Stress or pressure	10, 36,	13, 29,	35, 10, 36	35, 1, 14, 16	10, 15,	10, 15,	6, 35, 10	35, 24	6, 35, 36	36, 35, 21		35, 4, 15, 10	35, 33, 2,	9, 18, 3, 40	19, 3, 27		35, 39,	-	14, 24,		10, 35, 14	2, 36, 25	10, 36, 3,		37, 36, 4	10, 14, 36	10, 13,	6, 28, 25	3, 35	22, 2, 37	2, 33, 27, 18	1, 35, 16	11	2	35	19, 1, 35	2, 36, 37	35, 24	10, 14,
12	Shape	8, 10, 29, 40	15, 10,	29, 34, 5,	13, 14,	5, 34, 4, 10		14, 4, 15, 22	7, 2, 35	35, 15,	35, 10,	34, 15,		33, 1, 18, 4	30, 14,	14, 26, 9,		22, 14,	13, 15, 32	2, 6, 34, 14		4, 6, 2	14	35, 29, 3,		14, 10,	36, 22	10, 40, 16	28, 32, 1	32, 30, 40	22, 1, 2, 35	35, 1	1, 32, 17, 28	32, 15, 26	2, 13, 1	1, 15, 29	16, 29, 1,	15, 13, 39	15, 1, 32	17, 26,
13	Stability of the object's composition	21, 35, 2,	26, 39, 1,	13, 15, 1,	37	2, 11, 13	39	28, 10,	34, 28,	33, 15,	10, 35,	2, 35, 40	22, 1, 18, 4		17, 9, 15	13, 27,	39, 3, 35, 23	35, 1, 32	32, 3, 27, 16	13, 19	27, 4, 29, 18	32, 35,	14, 2, 39, 6	2, 14, 30, 40		35, 27	15, 32, 35		13	18	35, 24,	35, 40,	35, 19	32, 35, 30	2, 35, 10, 16	35, 30,	2, 35, 22, 26	35, 22,	1, 8, 35	23, 35,
14	Strength	1, 8, 40, 15	40, 26,	1, 15, 8, 35	15, 14,	3, 34, 40, 29	9, 40, 28	10, 15,	9, 14, 17, 15	8, 13, 26, 14	10, 18, 3,	10, 3, 18, 40	10, 30,	13, 17, 35		27, 3, 26		30, 10, 40	35, 19	19, 35, 10	35	10, 26,	35	35, 28,		29, 3, 28, 10	29, 10, 27	11, 3	3, 27, 16	3, 27	18, 35,	15, 35,	11, 3, 10, 32	32, 40,	27, 11, 3	15, 3, 32	2, 13, 25, 28	27, 3, 15, 40	15	29, 35,
15	Duration of action of moving object	19, 5, 34, 31	-	2, 19, 9	-	3, 17, 19	-	10, 2, 19, 30	-	3, 35, 5	19, 2, 16	19, 3, 27	14, 26,	13, 3, 35	27, 3, 10		-	19, 35, 39	2, 19, 4, 35	28, 6, 35, 18		19, 10,		28, 27, 3,	10	20, 10,	3, 35, 10, 40	11, 2, 13	3	3, 27, 16, 40	22, 15,	21, 39,	27, 1, 4	12, 27	29, 10, 27	1, 35, 13	10, 4, 29, 15	19, 29,	6, 10	35, 17,
16	Duration of action by stationary object	-	6, 27, 19, 16	-	1, 40, 35	-		-	35, 34, 38	-				39, 3, 35, 23		-		19, 18,		-		16		27, 16,	10	28, 20,	3, 35, 31	34, 27, 6,	10, 26, 24		17, 1, 40, 33	22	35, 10	1	1	2		25, 34, 6,	1	20, 10,
17	Temperature	36,22, 6, 38	22, 35, 32	15, 19, 9	15, 19, 9	3, 35, 39, 18	35, 38	34, 39,	35, 6, 4	2, 28, 36, 30	35, 10, 3,	35, 39,	14, 22,	1, 35, 32	10, 30,	19, 13, 39	19, 18,		32, 30,	19, 15, 3,		2, 14, 17, 25	21, 17,	21, 36,		35, 28,	3, 17, 30, 39	19, 35, 3,	32, 19, 24	24	22, 33,	22, 35, 2,	26, 27	26, 27	4, 10, 16	2, 18, 27	2, 17, 16	3, 27, 35, 31	26, 2, 19, 16	15, 28, 35
18	Illumination intensity	19, 1, 32	2, 35, 32	19, 32, 16		19, 32, 26		2, 13, 10		10, 13, 19	26, 19, 6		32, 30	32, 3, 27	35, 19	2, 19, 6		32, 35, 19		32, 1, 19	32, 35, 1,	32	13, 16, 1,	13, 1	1, 6	19, 1, 26, 17	1, 19		11, 15, 32	3, 32	15, 19	35, 19,	19, 35,	28, 26, 19	15, 17,	15, 1, 19	6, 32, 13	32, 15	2, 26, 10	2, 25, 16
19	Use of energy by moving object	12,18, 28,31	-	12, 28	-	15, 19, 25	-	35, 13, 18	-	8, 35, 35	16, 26,	23, 14, 25	12, 2, 29	19, 13,	5, 19, 9, 35	28, 35, 6,	-	19, 24, 3,	2, 15, 19		-	6, 19, 37, 18	12, 22,	35, 24,		35, 38,	34, 23,	19, 21,	3, 1, 32		1, 35, 6, 27	2, 35, 6	28, 26, 30	19, 35	1, 15, 17, 28	15, 17,	2, 29, 27, 28	35, 38	32, 2	12, 28, 35
20	Use of energy by stationary object	-	19, 9, 6, 27	-		-		-		-	36, 37			27, 4, 29, 18	35				19, 2, 35, 32	-				28, 27,			3, 35, 31	10, 36, 23			10, 2, 22, 37	19, 22, 18	1, 4					19, 35,		1, 6
21	Power	8, 36, 38, 31	19, 26,	1, 10, 35, 37		19, 38	17, 32,	35, 6, 38	30, 6, 25	15, 35, 2	26, 2, 36, 35	22, 10, 35	29, 14, 2,	35, 32,	26, 10, 28	19, 35,	16	2, 14, 17, 25	16, 6, 19	16, 6, 19, 37			10, 35, 38	28, 27,	10, 19	35, 20,	4, 34, 19	19, 24,	32, 15, 2	32, 2	19, 22,	2, 35, 18	26, 10, 34	26, 35, 10	35, 2, 10, 34	19, 17, 34	20, 19,	19, 35, 16	28, 2, 17	28, 35, 34
22	Loss of Energy	15, 6, 19, 28	19, 6, 18, 9	7, 2, 6, 13	6, 38, 7	15, 26,	17, 7, 30, 18	7, 18, 23	7	16, 35, 38	36, 38			14, 2, 39, 6	26			19, 38, 7	1, 13, 32, 15			3, 38		35, 27, 2,	19, 10	10, 18,	7, 18, 25	11, 10, 35	32		21, 22,	21, 35, 2,		35, 32, 1	2, 19		7, 23	35, 3, 15, 23	2	28, 10,
23	Loss of substance	35, 6, 23, 40	35, 6, 22, 32	14, 29,	10, 28,24	35, 2, 10, 31	10, 18,	1, 29, 30, 36	3, 39, 18, 31	10, 13,	14, 15,	3, 36, 37, 10	29, 35, 3,	2, 14, 30, 40	35, 28,	28, 27, 3,	27, 16,	21, 36,	1, 6, 13	35, 18,	28, 27,	28, 27,	35, 27, 2,			15, 18,	6, 3, 10, 24	10, 29,	16, 34,	35, 10,	33, 22,	10, 1, 34, 29	15, 34, 33	32, 28, 2,	2, 35, 34, 27	15, 10, 2	35, 10,	35, 18,	35, 10, 18	28, 35,
24	Loss of Information	10, 24, 35	10, 35, 5	1, 26	26	30, 26	30, 16		2, 22	26, 32						10	10		19			10, 19	19, 10			24, 26,	24, 28, 35	10, 28, 23			22, 10, 1	10, 21, 22	32	27, 22				35, 33	35	13, 23, 15
25	Loss of Time	10, 20,	10, 20,	15, 2, 29	30, 24,	26, 4, 5, 16	10, 35,	2, 5, 34, 10	35, 16,		10, 37,	37, 36,4	4, 10, 34, 17	35, 3, 22, 5	29, 3, 28, 18	20, 10,	28, 20,	35, 29,	1, 19, 26, 17	35, 38,	1	35, 20,	10, 5, 18, 32	35, 18,	24, 26,		35, 38,	10, 30, 4	24, 34,	24, 26,	35, 18, 34	35, 22,	35, 28,	4, 28, 10, 34	32, 1, 10	35, 28	6, 29	18, 28,	24, 28,	
26	Quantity of substance/the matter	35, 6, 18, 31	27, 26,	29, 14,		15, 14, 29	2, 18, 40, 4	15, 20, 29		35, 29,	35, 14, 3	10, 36,	35, 14	15, 2, 17, 40	14, 35,	3, 35, 10, 40	3, 35, 31	3, 17, 39		34, 29,	3, 35, 31	35	7, 18, 25	6, 3, 10, 24	24, 28, 35	35, 38,		18, 3, 28, 40	13, 2, 28	33, 30	35, 33,	3, 35, 40, 39	29, 1, 35, 27	35, 29,	2, 32, 10, 25	15, 3, 29	3, 13, 27, 10	3, 27, 29, 18	8, 35	13, 29, 3,
27	Reliability	3, 8, 10, 40	3, 10, 8, 28	15, 9, 14, 4	15, 29,	17, 10,	32, 35,	3, 10, 14, 24	2, 35, 24	21, 35,	8, 28, 10, 3	10, 24,	35, 1, 16, 11		11, 28	2, 35, 3, 25	34, 27, 6,	3, 35, 10	11, 32, 13	21, 11,	36, 23	21, 11,	10, 11, 35	10, 35,	10, 28	10, 30, 4	21, 28,		32, 3, 11, 23	11, 32, 1	27, 35, 2,	35, 2, 40, 26		27, 17, 40	1, 11	13, 35, 8,	13, 35, 1	27, 40, 28	11, 13, 27	1, 35, 29, 38
28	Measurement accuracy	32, 35,	28, 35,	28, 26, 5,	32, 28, 3,	26, 28,	26, 28,	32, 13, 6		28, 13,	32, 2	6, 28, 32	6, 28, 32	32, 35, 13	28, 6, 32	28, 6, 32	10, 26, 24	6, 19, 28, 24	6, 1, 32	3, 6, 32		3, 6, 32	26, 32, 27	10, 16,		24, 34,	2, 6, 32	5, 11, 1, 23			28, 24,	3, 33, 39, 10	6, 35, 25, 18	1, 13, 17, 34	1, 32, 13, 11	13, 35, 2	27, 35,	26, 24,	28, 2, 10, 34	10, 34,
29	Manufacturing precision	28, 32,	28, 35,	10, 28,	2, 32, 10	28, 33,	2, 29, 18, 36	32, 23, 2	25, 10, 35	10, 28, 32	28, 19,	3, 35	32, 30, 40	30, 18	3, 27	3, 27, 40		19, 26	3, 32	32, 2		32, 2	13, 32, 2	35, 31,		32, 26,	32, 30	11, 32, 1			26, 28,	4, 17, 34, 26		1, 32, 35, 23	25, 10		26, 2, 18		26, 28,	10, 18,
30	Object-affected harmful factors	22, 21,	2, 22, 13, 24	17, 1, 39, 4	1, 18	22, 1, 33, 28	27, 2, 39, 35	22, 23,	34, 39,	21, 22,	13, 35,	22, 2, 37	22, 1, 3, 35	35, 24,	18, 35,	22, 15,	17, 1, 40, 33	22, 33,	1, 19, 32, 13	1, 24, 6, 27	10, 2, 22, 37	19, 22,	21, 22,	33, 22,	22, 10, 2	35, 18, 34	35, 33,	27, 24, 2,	28, 33,	26, 28,			24, 35, 2	2, 25, 28, 39	35, 10, 2	35, 11,	22, 19,	22, 19,	33, 3, 34	22, 35,
31	Object-generated harmful factors	19, 22,	35, 22, 1,	17, 15,		17, 2, 18, 39	22, 1, 40	17, 2, 40	30, 18,	35, 28, 3,	35, 28, 1,	2, 33, 27, 18	35, 1	35, 40,	15, 35,	15, 22,	21, 39,	22, 35, 2,	19, 24,	2, 35, 6	19, 22, 18	2, 35, 18	21, 35, 2,	10, 1, 34	10, 21, 29	1, 22	3, 24, 39, 1	24, 2, 40, 39	3, 33, 26	4, 17, 34, 26							19, 1, 31	2, 21, 27, 1	2	22, 35,
32	Ease of manufacture	28, 29,	1, 27, 36, 13	1, 29, 13, 17	15, 17, 27	13, 1, 26, 12	16, 40	13, 29, 1,	35	35, 13, 8,	35, 12	35, 19, 1,	1, 28, 13, 27	11, 13, 1	1, 3, 10, 32	27, 1, 4	35, 16	27, 26, 18	28, 24,	28, 26,	1, 4	27, 1, 12, 24	19, 35	15, 34, 33	32, 24,	35, 28,	35, 23, 1,		1, 35, 12, 18		24, 2			2, 5, 13, 16	35, 1, 11, 9	2, 13, 15	27, 26, 1	6, 28, 11, 1	8, 28, 1	35, 1, 10, 28
33	Ease of operation	25, 2, 13, 15	6, 13, 1, 25	1, 17, 13, 12		1, 17, 13, 16	18, 16,	1, 16, 35, 15	4, 18, 39, 31	18, 13, 34	28, 13 35	2, 32, 12	15, 34,	32, 35, 30	32, 40, 3,	29, 3, 8, 25	1, 16, 25	26, 27, 13	13, 17, 1,	1, 13, 24		35, 34, 2,	2, 19, 13	28, 32, 2,	4, 10, 27, 22	4, 28, 10, 34	12, 35	17, 27, 8,	25, 13, 2,	1, 32, 35, 23	2, 25, 28, 39		2, 5, 12		12, 26, 1,	15, 34, 1,	32, 26,		1, 34, 12, 3	15, 1, 28
34	Ease of repair	2, 27 35, 11	2, 27, 35, 11	1, 28, 10, 25	3, 18, 31	15, 13, 32	16, 25	25, 2, 35, 11	1	34, 9	1, 11, 10	13	1, 13, 2, 4	2, 35	11, 1, 2, 9	11, 29,	1	4, 10	15, 1, 13	15, 1, 28, 16		15, 10,	15, 1, 32, 19	2, 35, 34, 27		32, 1, 10, 25	2, 28, 10, 25	11, 10, 1,	10, 2, 13	25, 10	35, 10, 2,		1, 35, 11, 10	1, 12, 26, 15		7, 1, 4, 16	35, 1, 13, 11		34, 35, 7,	1, 32, 10
35	Adaptability or versatility	1, 6, 15, 8	19, 15,	35, 1, 29, 2	1, 35, 16	35, 30,	15, 16	15, 35, 29		35, 10, 14	15, 17, 20	35, 16	15, 37, 1,	35, 30, 14	35, 3, 32, 6	13, 1, 35	2, 16	27, 2, 3, 35	6, 22, 26, 1	19, 35,		19, 1, 29	18, 15, 1	15, 10, 2,		35, 28	3, 35, 15	35, 13, 8,	35, 5, 1, 10		35, 11,		1, 13, 31	15, 34, 1,	1, 16, 7, 4		15, 29,	1	27, 34, 35	35, 28, 6,
36	Device complexity	26, 30,	2, 26, 35, 39	1, 19, 26, 24	26	14, 1, 13, 16	6, 36	34, 26, 6	1, 16	34, 10, 28	26, 16	19, 1, 35	29, 13,	2, 22, 17, 19	2, 13, 28	10, 4, 28, 15		2, 17, 13	24, 17, 13	27, 2, 29, 28		20, 19,	10, 35,	35, 10,		6, 29	13, 3, 27, 10	13, 35, 1	2, 26, 10, 34	26, 24, 32	22, 19,	19, 1	27, 26, 1,	27, 9, 26, 24	1, 13	29, 15,		15, 10,	15, 1, 24	12, 17, 28
37	Difficulty of detecting and measuring	27, 26,	6, 13, 28, 1	16, 17,	26	2, 13, 18, 17	2, 39, 30, 16	29, 1, 4, 16	2, 18, 26, 31	3, 4, 16, 35	30, 28,	35, 36,	27, 13, 1,	11, 22,	27, 3, 15, 28	19, 29,	25, 34, 6,	3, 27, 35, 16	2, 24, 26	35, 38	19, 35, 16	18, 1, 16, 10	35, 3, 15, 19	1, 18, 10, 24	35, 33,	18, 28,	3, 27, 29, 18	27, 40,	26, 24,		22, 19,	2, 21	5, 28, 11, 29	2, 5	12, 26	1, 15	15, 10,		34, 21	35, 18
38	Extent of automation	28, 26,	28, 26,	14, 13,	23	17, 14, 13		35, 13, 16		28, 10	2, 35	13, 35	15, 32, 1,	18, 1	25, 13	6, 9		26, 2, 19	8, 32, 19	2, 32, 13		28, 2, 27	23, 28	35, 10,	35, 33	24, 28,	35, 13	11, 27, 32	28, 26,	28, 26,	2, 33	2	1, 26, 13	1, 12, 34, 3	1, 35, 13	27, 4, 1, 35	15, 24, 10	34, 27, 25		5, 12, 35, 26

Appendix B

40 TRIZ Inventive Principles

1. Segmentation
2. Separation
3. Local quality
4. Asymmetry
5. Merging
6. Universality
7. Nesting
8. Anti-weight
9. Preliminary counteraction
10. Preliminary action
11. Beforehand compensation
12. Equi-potentiality
13. Do it in reverse ('The other way round')
14. Curvature
15. Dynamics
16. Partial or excessive actions
17. Change in dimension
18. Mechanical vibration

19. Periodic action
20. Continuity of useful action
21. Expediting (Skipping)
22. Convert harm to benefit ("Blessing in disguise")
23. Feedback
24. Intermediary
25. Self-service
26. Copying
27. Cheap disposable objects
28. Mechanical substitutions
29. Pneumatics and hydraulics
30. Flexible shells and thin films
31. Porous materials
32. Change in optical properties
33. Homogeneity
34. Discarding and recovering
35. Parameter changes
36. Phase transitions
37. Thermal expansion
38. Strong oxidants
39. Inert atmosphere
40. Composite materials

Appendix C

Some Useful Orthogonal Arrays

Table C.1

The orthogonal array is denoted as $L_a\ (b^c)$ where:

- L = Latin square
- a = The number of test trials
- b = The number of levels for each column
- c = The number of columns in the array

TWO-LEVEL ORTHOGONAL ARRAYS

Table C.2

$L_4\ (2^3)$ Orthogonal Array

No.	*1*	*2*	*3*
1	1	1	1
2	1	2	2
3	2	1	2
4	2	2	1

Table C.2 (*Continued*)

$L_8\ (2^7)$ Orthogonal Array

No.	*1*	*2*	*3*	*4*	*5*	*6*	*7*
1	1	1	1	1	1	1	1
2	1	1	1	2	2	2	2
3	1	2	2	1	1	2	2
4	1	2	2	2	2	1	1
5	2	1	2	1	2	1	2
6	2	1	2	2	1	2	1
7	2	2	1	1	2	2	1
8	2	2	1	2	1	1	2

Table C.3 $L_{12}\ (2^{11})$ Orthogonal Array

No.	*1*	*2*	*3*	*4*	*5*	*6*	*7*	*8*	*9*	*10*	*11*
1	1	1	1	1	1	1	1	1	1	1	1
2	1	1	1	1	1	2	2	2	2	2	2
3	1	1	2	2	2	1	1	1	2	2	2
4	1	2	1	2	2	1	2	2	1	1	2
5	1	2	2	1	2	2	1	2	1	2	1
6	1	2	2	2	1	2	2	1	2	1	1
7	2	1	2	2	1	1	2	2	1	2	1
8	2	1	2	1	2	2	2	1	1	1	2
9	2	1	1	2	2	2	1	2	2	1	1
10	2	2	2	1	1	1	1	2	2	1	2
11	2	2	1	2	1	2	1	1	1	2	2
12	2	2	1	1	2	1	2	1	2	2	1

- This is a special orthogonal array where interactions are distributed to all columns, more or less uniformly.
- Conclusions regarding main effects are more robust against confounding of interactions.

Table C.4 L_{16} (2^{15}) Orthogonal Array

No.	*1*	*2*	*3*	*4*	*5*	*6*	*7*	*8*	*9*	*10*	*11*	*12*	*13*	*14*	*15*
1	1	1	1	1	1	1	1	1	1	1	1	1	1	1	1
2	1	1	1	1	1	1	1	2	2	2	2	2	2	2	2
3	1	1	1	2	2	2	2	1	1	1	1	2	2	2	2
4	1	1	1	2	2	2	2	2	2	2	2	1	1	1	1
5	1	2	2	1	1	2	2	1	1	2	2	1	1	2	2
6	1	2	2	1	1	2	2	2	2	1	1	2	2	1	1
7	1	2	2	2	2	1	1	1	1	2	2	2	2	1	1
8	1	2	2	2	2	1	1	2	2	1	1	1	1	2	2
9	2	1	2	1	2	1	2	1	2	1	2	1	2	1	2
10	2	1	2	1	2	1	2	2	1	2	1	2	1	2	1
11	2	1	2	2	1	2	1	1	2	1	2	2	1	2	1
12	2	1	2	2	1	2	1	2	1	2	1	1	2	1	2
13	2	2	1	1	2	2	1	1	2	2	1	1	2	2	1
14	2	2	1	1	2	2	1	2	1	1	2	2	1	1	2
15	2	2	1	2	1	1	2	1	2	2	1	2	1	1	2
16	2	2	1	2	1	1	2	2	1	1	2	1	2	2	1

Table C.5 L_{32} (2^{31}) Orthogonal Array

No.	*1*	*2*	*3*	*4*	*5*	*6*	*7*	*8*	*9*	*10*	*11*	*12*	*13*	*14*	*15*	*16*	*17*	*18*	*19*	*20*	*21*	*22*	*23*	*24*	*25*	*26*	*27*	*28*	*29*	*30*	*31*
1	1	1	1	1	1	1	1	1	1	1	1	1	1	1	1	1	1	1	1	1	1	1	1	1	1	1	1	1	1	1	1
2	1	1	1	1	1	1	1	1	1	1	1	1	1	1	1	2	2	2	2	2	2	2	2	2	2	2	2	2	2	2	2
3	1	1	1	1	1	1	1	2	2	2	2	2	2	2	2	1	1	1	1	1	1	1	1	2	2	2	2	2	2	2	2
4	1	1	1	1	1	1	1	2	2	2	2	2	2	2	2	2	2	2	2	2	2	2	2	1	1	1	1	1	1	1	1
5	1	1	1	2	2	2	2	1	1	1	1	2	2	2	2	1	1	1	1	2	2	2	2	1	1	1	1	2	2	2	2
6	1	1	1	2	2	2	2	1	1	1	1	2	2	2	2	2	2	2	2	1	1	1	1	2	2	2	2	1	1	1	1
7	1	1	1	2	2	2	2	2	2	2	2	1	1	1	1	1	1	1	1	2	2	2	2	2	2	2	2	1	1	1	1
8	1	1	1	2	2	2	2	2	2	2	2	1	1	1	1	2	2	2	2	1	1	1	1	1	1	1	1	2	2	2	2
9	1	2	2	1	1	2	2	1	1	2	2	1	1	2	2	1	1	2	2	1	1	2	2	1	1	2	2	1	1	2	2
10	1	2	2	1	1	2	2	1	1	2	2	1	1	2	2	2	2	1	1	2	2	1	1	2	2	1	1	2	2	1	1
11	1	2	2	1	1	2	2	2	2	1	1	2	2	1	1	1	1	2	2	1	1	2	2	2	2	1	1	2	2	1	1
12	1	2	2	1	1	2	2	2	2	1	1	2	2	1	1	2	2	1	1	2	2	1	1	1	1	2	2	1	1	2	2
13	1	2	2	2	2	1	1	1	1	2	2	2	2	1	1	1	1	2	2	2	2	1	1	1	1	2	2	2	2	1	1
14	1	2	2	2	2	1	1	1	1	2	2	2	2	1	1	2	2	1	1	1	1	2	2	2	2	1	1	1	1	2	2
15	1	2	2	2	2	1	1	2	2	1	1	1	1	2	2	1	1	2	2	2	2	1	1	2	2	1	1	1	1	2	2
16	1	2	2	2	2	1	1	2	2	1	1	1	1	2	2	2	2	1	1	1	1	2	2	1	1	2	2	2	2	1	1
17	2	1	2	1	2	1	2	1	2	1	2	1	2	1	2	1	2	1	2	1	2	1	2	1	2	1	2	1	2	1	2
18	2	1	2	1	2	1	2	1	2	1	2	1	2	1	2	2	1	2	1	2	1	2	1	2	1	2	1	2	1	2	1

19	2	1	2	1	2	1	2	2	1	2	1	2	1	2	1	1	2	1	2	1	2	1	2	2	1	2	1	2	1	2	1
20	2	1	2	1	2	1	2	2	1	2	1	2	1	2	1	2	1	2	1	2	1	2	1	1	2	1	2	1	2	1	2
21	2	1	2	2	1	2	1	1	2	1	2	2	1	2	1	1	2	1	2	2	1	2	1	1	2	1	2	2	1	2	1
22	2	1	2	2	1	2	1	1	2	1	2	2	1	2	1	2	1	2	1	1	2	1	2	2	1	2	1	1	2	1	2
23	2	1	2	2	1	2	1	2	1	2	1	1	2	1	2	1	2	1	2	2	1	2	1	2	1	2	1	1	2	1	2
24	2	1	2	2	1	2	1	2	1	2	1	1	2	1	2	2	1	2	1	1	2	1	2	1	2	1	2	2	1	2	1
25	2	2	1	1	2	2	1	1	2	2	1	1	2	2	1	1	2	2	1	1	2	2	1	1	2	2	1	1	2	2	1
26	2	2	1	1	2	2	1	1	2	2	1	1	2	2	1	2	1	1	2	2	1	1	2	2	1	1	2	2	1	1	2
27	2	2	1	1	2	2	1	2	1	1	2	2	1	1	2	1	2	2	1	1	2	2	1	2	1	1	2	2	1	1	2
28	2	2	1	1	2	2	1	2	1	1	2	2	1	1	2	2	1	1	2	2	1	1	2	1	2	2	1	1	2	2	1
29	2	2	1	2	1	1	2	1	2	2	1	2	1	1	2	1	2	2	1	2	1	1	2	1	2	2	1	2	1	1	2
30	2	2	1	2	1	1	2	1	2	2	1	2	1	1	2	2	1	1	2	1	2	2	1	2	1	1	2	1	2	2	1
31	2	2	1	2	1	1	2	2	1	1	2	1	2	2	1	1	2	2	1	2	1	1	2	2	1	1	2	1	2	2	1
32	2	2	1	2	1	1	2	2	1	1	2	1	2	2	1	2	1	1	2	1	2	2	1	1	2	2	1	2	1	1	2

THREE-LEVEL ORTHOGONAL ARRAYS

Table C.6 L_9 (3^4) Orthogonal Array

No.	*1*	*2*	*3*	*4*
1	1	1	1	1
2	1	2	2	2
3	1	3	3	3
4	2	1	2	3
5	2	2	3	1
6	2	3	1	2
7	3	1	3	2
8	3	2	1	3
9	3	3	2	1

Table C.7 L_{18} $(2^1 \times 3^7)$ Orthogonal Array

No.	*1*	*2*	*3*	*4*	*5*	*6*	*7*	*8*
1	1	1	1	1	1	1	1	1
2	1	1	2	2	2	2	2	2
3	1	1	3	3	3	3	3	3
4	1	2	1	1	2	2	3	3
5	1	2	2	2	3	3	1	1
6	1	2	3	3	1	1	2	2
7	1	3	1	2	1	3	2	3
8	1	3	2	3	2	1	3	1
9	1	3	3	1	3	2	1	2
10	2	1	1	3	3	2	2	1
11	2	1	2	1	1	3	3	2
12	2	1	3	2	2	1	1	3
13	2	2	1	2	3	1	3	2
14	2	2	2	3	1	2	1	3
15	2	2	3	1	2	3	2	1
16	2	3	1	3	2	3	1	2
17	2	3	2	1	3	1	2	3
18	2	3	3	2	1	2	3	1

- This is a special orthogonal array where interactions are distributed to all columns, more or less uniformly.
- Conclusions regarding main effects are more robust against confounding of interactions.
- Highly recommended for robust optimization.
- This array is very popular for robust optimization using computer simulations and system testing.

Table C.8 L_{27} (3^{13}) Orthogonal Array

No.	*1*	*2*	*3*	*4*	*5*	*6*	*7*	*8*	*9*	*10*	*11*	*12*	*13*
1	1	1	1	1	1	1	1	1	1	1	1	1	1
2	1	1	1	1	2	2	2	2	2	2	2	2	2
3	1	1	1	1	3	3	3	3	3	3	3	3	3
4	1	2	2	2	1	1	1	2	2	2	3	3	3
5	1	2	2	2	2	2	2	3	3	3	1	1	1
6	1	2	2	2	3	3	3	1	1	1	2	2	2
7	1	3	3	3	1	1	1	3	3	3	2	2	2
8	1	3	3	3	2	2	2	1	1	1	3	3	3
9	1	3	3	3	3	3	3	2	2	2	1	1	1
10	2	1	2	3	1	2	3	1	2	3	1	2	3
11	2	1	2	3	2	3	1	2	3	1	2	3	1
12	2	1	2	3	3	1	2	3	1	2	3	1	2
13	2	2	3	1	1	2	3	2	3	1	3	1	2
14	2	2	3	1	2	3	1	3	1	2	1	2	3
15	2	2	3	1	3	1	2	1	2	3	2	3	1
16	2	3	1	2	1	2	3	3	1	2	2	3	1
17	2	3	1	2	2	3	1	1	2	3	3	1	2
18	2	3	1	2	3	1	2	2	3	1	1	2	3
19	3	1	3	2	1	3	2	1	3	2	1	3	2
20	3	1	3	2	2	1	3	2	1	3	2	1	3
21	3	1	3	2	3	2	1	3	2	1	3	2	1
22	3	2	1	3	1	3	2	2	1	3	3	2	1
23	3	2	1	3	2	1	3	3	2	1	1	3	2
24	3	2	1	3	3	2	1	1	3	2	2	1	3
25	3	3	2	1	1	3	2	3	2	1	2	1	3
26	3	3	2	1	2	1	3	1	3	2	3	2	1
27	3	3	2	1	3	2	1	2	1	3	1	3	2

Table C.9 L_{36} $(2^{11} \times 3^{12})$ Orthogonal Array

No.	*1*	*2*	*3*	*4*	*5*	*6*	*7*	*8*	*9*	*10*	*11*	*12*	*13*	*14*	*15*	*16*	*17*	*18*	*19*	*20*	*21*	*22*	*23*
1	1	1	1	1	1	1	1	1	1	1	1	1	1	1	1	1	1	1	1	1	1	1	1
2	1	1	1	1	1	1	1	1	1	1	1	2	2	2	2	2	2	2	2	2	2	2	2
3	1	1	1	1	1	1	1	1	1	1	1	3	3	3	3	3	3	3	3	3	3	3	3
4	1	1	1	1	1	2	2	2	2	2	2	1	1	1	1	2	2	2	2	3	3	3	3
5	1	1	1	1	1	2	2	2	2	2	2	2	2	2	2	3	3	3	3	1	1	1	1
6	1	1	1	1	1	2	2	2	2	2	2	3	3	3	3	1	1	1	1	2	2	2	2
7	1	1	2	2	2	1	1	1	2	2	2	1	1	2	3	1	2	3	3	1	2	2	3
8	1	1	2	2	2	1	1	1	2	2	2	2	2	3	1	2	3	1	1	2	3	3	1
9	1	1	2	2	2	1	1	1	2	2	2	3	3	1	2	3	1	2	2	3	1	1	2
10	1	2	1	2	2	1	2	2	1	1	2	1	1	3	2	1	3	2	3	2	1	3	2
11	1	2	1	2	2	1	2	2	1	1	2	2	2	1	3	2	1	3	1	3	2	1	3
12	1	2	1	2	2	1	2	2	1	1	2	3	3	2	1	3	2	1	2	1	3	2	1
13	1	2	2	1	2	2	1	2	1	2	1	1	2	3	1	3	2	1	3	3	2	1	2
14	1	2	2	1	2	2	1	2	1	2	1	2	3	1	2	1	3	2	1	1	3	2	3
15	1	2	2	1	2	2	1	2	1	2	1	3	1	2	3	2	1	3	2	2	1	3	1
16	1	2	2	2	1	2	2	1	2	1	1	1	2	3	2	1	1	3	2	3	3	2	1
17	1	2	2	2	1	2	2	1	2	1	1	2	3	1	3	2	2	1	3	1	1	3	2
18	1	2	2	2	1	2	2	1	2	1	1	3	1	2	1	3	3	2	1	2	2	1	3
19	2	1	2	2	1	1	2	2	1	2	1	1	2	1	3	3	3	1	2	2	1	2	3
20	2	1	2	2	1	1	2	2	1	2	1	2	3	2	1	1	1	2	3	3	2	3	1
21	2	1	2	2	1	1	2	2	1	2	1	3	1	3	2	2	2	3	1	1	3	1	2
22	2	1	2	1	2	2	2	1	1	1	2	1	2	2	3	3	1	2	1	1	3	3	2
23	2	1	2	1	2	2	2	1	1	1	2	2	3	3	1	1	2	3	2	2	1	1	3
24	2	1	2	1	2	2	2	1	1	1	2	3	1	1	2	2	3	1	3	3	2	2	1
25	2	1	1	2	2	2	1	2	2	1	1	1	3	2	1	2	3	3	1	3	1	2	2
26	2	1	1	2	2	2	1	2	2	1	1	2	1	3	2	3	1	1	2	1	2	3	3
27	2	1	1	2	2	2	1	2	2	1	1	3	2	1	3	1	2	2	3	2	3	1	1
28	2	2	2	1	1	1	1	2	2	1	2	1	3	2	2	2	1	1	3	2	3	1	3
29	2	2	2	1	1	1	1	2	2	1	2	2	1	3	3	3	2	2	1	3	1	2	1
30	2	2	2	1	1	1	1	2	2	1	2	3	2	1	1	1	3	3	2	1	2	3	2
31	2	2	1	2	1	2	1	1	1	2	2	1	3	3	3	2	3	2	2	1	2	1	1
32	2	2	1	2	1	2	1	1	1	2	2	2	1	1	1	3	1	3	3	2	3	2	2
33	2	2	1	2	1	2	1	1	1	2	2	3	2	2	2	1	2	1	1	3	1	3	3
34	2	2	1	1	2	1	2	1	2	2	1	1	3	1	2	3	2	3	1	2	2	3	1
35	2	2	1	1	2	1	2	1	2	2	1	2	1	2	3	1	3	1	2	3	3	1	2
36	2	2	1	1	2	1	2	1	2	2	1	3	2	3	1	2	1	2	3	1	1	2	3

Appendix D

Equations for Signal-to-noise (S/N) Ratios

NONDYNAMIC SIGNAL-TO-NOISE RATIO

- Nominal-the-best
- Smaller-the-better
- Larger-the-better

Nominal-the-Best Type (Type I)

Let data points be $y_1\ y_2 \ldots y_n$

$$T = \text{Sum of data} = \sum_{i=1}^{n} y_i$$

$$S_m = \text{Sum of squares due to mean} = \frac{T^2}{n}$$

$$V_e = \text{Mean square (variance)} = \sigma_{n-1}^2 = \sum_{i=1}^{n} \frac{(y_i - \bar{y})^2}{n-1}$$

$$\text{S/N} = \eta_{\text{dB}} = 10\text{Log}\left[\frac{\frac{1}{n}(S_m - V_e)}{V_e}\right]$$

Nominal-the-Best Type (Type II)

Let data points be $y_1\ y_2 \ldots y_n$

$$\text{S/N} = \eta_{\text{dB}} = 10\text{Log}\left[\frac{\frac{1}{n}(S_m - V_e)}{V_e}\right] = 10\text{Log}\left[\frac{\frac{1}{n}\left(\frac{T^2}{n} - \sigma_{n-1}^2\right)}{\sigma_{n-1}^2}\right]$$

$$= 10\text{Log}\left[\frac{\frac{T^2}{n^2}}{\sigma_{n-1}^2}\right]$$

$$= 10\text{Log}\left[\frac{y^{-2}}{\sigma_{n-1}^2}\right]$$

We can also show that nominal-the-best S/N ratio as:

$$\text{S/N} = \eta_{\text{dB}} = 10\text{Log}\left[\frac{1}{V_e}\right] = 10\text{Log}\left[\frac{1}{\sigma_{n-1}^2}\right]$$

In this equation, the error variance (V_e) is an unbiased estimate of σ^2.

*Note:*The higher the S/N becomes, the smaller the variability is. Maximizing this S/N is equivalent to minimizing standard deviation or variation.

Smaller-the-Better Type

Let data points be $y_1\ y_2 \ldots y_n$

$$\text{S/N} = \eta_{\text{dB}} = 10\text{Log}\left[\frac{1}{\frac{1}{n}\sum_{i=1}^{n} y_i^2}\right] = 10\text{Log}\left[\frac{1}{\bar{y}^2 + \sigma^2}\right]$$

Note: Maximizing this S/N is to minimize the mean and standard deviation.

Larger-the-Better Type

Let data points be $y_1\, y_2 \ldots y_n$

$$\text{S/N} = \eta_{\text{dB}} = 10\text{Log}\left[\frac{1}{\frac{1}{n}\sum_{i=1}^{n}\frac{1}{Y_i^2}}\right] = -10\text{Log}\left(\frac{1}{n}\sum_{i=1}^{n}\frac{1}{y_i^2}\right)$$

Maximizing this S/N is to maximize the mean and to minimize standard deviation.

DYNAMIC S/N RATIOS

Let the data set from outer array (sample size, $n = 16$) be as follows.

Table D.1

	M_1	M_2	M_3	M_4
N_1Q_1	Y_1	Y_2	Y_3	Y_4
N_1Q_2	Y_5	Y_6	Y_7	Y_8
N_2Q_1	Y_9	Y_{10}	Y_{11}	Y_{12}
N_2Q_2	Y_{13}	Y_{14}	Y_{15}	Y_{16}
Total	Y_1	Y_2	Y_3	Y_4

n = # of data points, sample size = 16

$r = M_1^2 + M_2^2 + M_3^2 + M_4^2$

r_0 = # of data in M_i = 4

Y_i = Total sum of data from M_i

$$\text{Total sum of squares } S_T = \sum_{i=1}^{n} = y_i^2$$

$$\text{Sum of squares due to slope } (\beta),\ S_\beta = \frac{1}{r \times r_0}[Y_1M_1 + Y_2M_2 + Y_3M_3 + Y_4M_4]^2$$

$$\text{Error sum of squares, } S_e = S_T - S_\beta$$

$$\text{Also error variance, } V_e = \frac{S_e}{n-1}$$

$$\text{S/N} = \eta_{\text{dB}} = 10\text{Log}\left[\frac{\frac{1}{r \times r_0}(S_\beta - V_e)}{V_e}\right]$$

Note: $\frac{1}{r \times r_0}(S_\beta - V_e)$ is an unbiased estimate of β^2

V_e is an unbiased estimate of mean square.

$$\beta = \sqrt{\frac{1}{r \times r_0}(S_\beta - V_e)} \cong \frac{1}{r \times r_0}(Y_1M_1 + Y_2M_2 + Y_3M_3 + Y_4M_4)$$

Appendix E

Related Topics of Matrix Theory

WHAT IS MATRIX?

A matrix is an array of elements arranged in rows and columns. Matrix manipulations play a significant role in multivariate analysis or pattern analysis. If a matrix A has m rows and n columns, then we say that matrix A is of size $m \times n$. An example of 3×4 matrix is shown below.

$$\mathrm{A} = \begin{bmatrix} a_{11} & a_{12} & a_{13} & a_{14} \\ a_{21} & a_{22} & a_{23} & a_{24} \\ a_{31} & a_{32} & a_{33} & a_{34} \end{bmatrix}$$

TRANSPOSE OF A MATRIX

If the rows and columns of a matrix A are interchanged, the resultant matrix is called transpose of matrix A and is denoted by A^T or A'. If A is of size $m \times n$ then A^T is of the size $n \times m$. The transpose of A is a 3×4 matrix and is shown below

$$A^T \text{ or } A' = \begin{bmatrix} a_{11} & a_{21} & a_{31} \\ a_{12} & a_{22} & a_{32} \\ a_{13} & a_{23} & a_{33} \\ a_{14} & a_{24} & a_{34} \end{bmatrix}$$

SQUARE MATRIX

If the number of rows and columns of a matrix is the same, then that matrix is called a square matrix.

DETERMINANT OF A MATRIX

The determinant is a characteristic number associated with a square matrix. The importance of the determinant can be realized when solving a system of linear equations using matrix algebra. The solution to the system of equations contains an inverse matrix term, which is obtained by dividing the adjoint matrix by the determinant. If the determinant is zero, then the solution does not exist.

Let us consider a 2×2 matrix:

$$A = \begin{bmatrix} a_{11} & a_{12} \\ a_{21} & a_{22} \end{bmatrix}$$

The determinant of this matrix is $a_{11} a_{22} - a_{12} a_{21}$.

Now let us consider a 3×3 matrix:

$$A = \begin{bmatrix} a_{11} & a_{12} & a_{13} \\ a_{21} & a_{22} & a_{23} \\ a_{31} & a_{32} & a_{33} \end{bmatrix}$$

The determinant of A can be calculated as:

$$\text{det. A} = a_{11}A_{11} + a_{12}A_{12} + a_{13}A_{13}$$

where,

$$A_{11} = (a_{22}\, a_{33} - a_{23}a_{32});$$
$$A_{12} = -\,(a_{21}\, a_{33} - a_{23}a_{31});$$
$$A_{13} = (a_{21}a_{32} - a_{22}a_{31})$$

are called cofactors of the elements a_{11}, a_{12}, and a_{13} of matrix A, respectively. The cofactors can be computed from submatrices obtained by deleting the rows and columns passing through the respective elements. Along a row or a column, the cofactors will have alternate plus and minus signs, with the first cofactor having a positive sign.

The previous equation for the determinant is obtained by using the elements of the first row and their cofactors. The same value of determinant can be obtained by using other rows or any column of the matrix with corresponding cofactors. In general, the determinant of a $n \times n$ square matrix can be written as:

$$\text{det. A} = a_{i1}A_{i1} + a_{i2}A_{i2} + \cdots + a_{in}A_{in} \quad \text{along any row } i, \quad \text{where } i = 1, 2, \ldots, n$$

or

$$\text{det. A} = a_{1j}A_{1j} + a_{2j}A_{2j} + \cdots + a_{nj}A_{nj} \quad \text{along any row } j, \quad \text{where } j = 1, 2, \ldots, n$$

COFACTOR

It is clear that the cofactor of A_{ij} of an element a_{ij} is the factor remaining after the element a_{ij} is factored out. The method of computing the cofactors was explained for a 3×3 matrix. Along a row or a column, the cofactors will have alternate signs of positive and negative, with the first cofactor having a positive sign.

ADJOINT MATRIX OF A SQUARE MATRIX

The adjoint of a square matrix A is obtained by replacing each element of A with its own cofactor and transposing the result.

Let us again consider a 3×3 matrix:

$$A = \begin{bmatrix} a_{11} & a_{12} & a_{13} \\ a_{21} & a_{22} & a_{23} \\ a_{31} & a_{32} & a_{33} \end{bmatrix}$$

The cofactor matrix containing cofactors (A_{ij}) of the elements of this matrix can be written as

$$A = \begin{bmatrix} A_{11} & A_{12} & A_{13} \\ A_{21} & A_{22} & A_{23} \\ A_{31} & A_{32} & A_{33} \end{bmatrix}$$

The adjoint of the matrix A, which is obtained by transposing the cofactor matrix, can be written as

$$\text{Adj. } A = \begin{bmatrix} A_{11} & A_{21} & A_{31} \\ A_{12} & A_{22} & A_{32} \\ A_{13} & A_{23} & A_{33} \end{bmatrix}$$

INVERSE MATRIX

The inverse of matrix A (denoted as A^{-1}) can be obtained by dividing the elements of its adjoint by the determinant. It should be noted that $A\,A^{-1} = A^{-1}\,A = I$, where I is identity matrix with all on-diagonal elements as 1 and off-diagonal elements as 0.

SINGULAR AND NONSINGULAR MATRICES

If the determinant of a square matrix is zero, then it is called a singular matrix. Otherwise, the matrix is known as nonsingular.

SOME OTHER DEFINITIONS

Correlation coefficient – The measure of linear association between the variables X_1 and X_2. This value lies between -1 and $+1$.

Correlation matrix – The matrix that gives correlation coefficients between the variables.

Standardized distance – Distance of an observation from the mean in terms of standard deviations.

Standardized variables – Variables obtained after subtracting the mean from the original variables and dividing the subtracted quantity by standard deviation.

Degrees of freedom – The number of independent parameters associated with an entity. These entities could be a matrix experiment, or a factor, or a sum of squares.

Normal distribution – The most commonly used distribution in statistics. This distribution is also known as Gaussian distribution. It is a bell-shaped curve and is symmetric about mean. The distribution is specified by two parameters, mean and standard deviation.

Index